The Designer's Guide to High-Purity Oscillators

The Designer's Guide Book Series

Consulting Editor: Kenneth S. Kundert

Books in the Series:

The Designer's Guide to Verilog-AMS
ISBN: 1-4020-8044-1

The Designer's Guide to SPICE and Spectre*
ISBN: 0-7923-9571-9

Emad Hegazi
Jacob Rael
Asad Abidi

The Designer's Guide to High-Purity Oscillators

Kluwer Academic Publishers

Library of Congress Cataloging-in-Publication Data

A C.I.P. Catalogue record for this book is available
from the Library of Congress.

ISBN 1-4020-7666-5 e-ISBN 0-387- 23294-X Printed on acid-free paper.

Printed in the United States of America.

9 8 7 6 5 4 3 2 1 SPIN 11053576

springeronline.com

Contents

Preface ix

Chapter 1 Basics of LC Oscillators 1

1 Introduction 1
2 The Mathematical Oscillator 1
3 Additive White Noise in LC Oscillators 3
4 The Linear Oscillator 3
4.1 Warning 3
4.2 Linear System Theory Applied to Oscillators 4
5 Linear Oscillator Noise Analysis 6
6 How Is This Book Different? 8

Chapter 2 Oscillator Purity Fundamentals 11

1 Introduction 11
2 Timing Jitter 12
3 Recognizing Phase Noise 17
4 Single Sideband Contains AM and PM 19
5 Phase Noise 19
6 Oscillator Phase Noise Models: Post-Leeson 24
6.1 Hajimiri's Model 25
6.2 Demir's Model 28
6.3 A Mechanistic Physical Model for LC Oscillators 31

Chapter 3 Current Biased Oscillator 35

1 Steady-State Operation 35
2 Linear Analysis of Differential Oscillator 37
3 Thermally Induced Phase Noise 39
3.1 Resonator Noise 44
3.2 Differential Pair Noise 50
3.3 Tail Current Noise 57
3.4 Proving Leeson's Hypothesis 62
4 Validation of Thermal Noise Analysis 63

Chapter 4 Colpitts Oscillator 67

1 Introduction 67
2 Steady-State 67

3 Phase Noise Analysis 72
3.1 Noise Sources 72
3.2 Noise in the Resistor 74
3.3 Noise of the Current Source and the Transistor 78
3.4 Noise Factor of Colpitts Oscillator 78
4 Conclusions 80

Chapter 5 Design for Low Thermal Phase Noise 83
1 Introduction 83
1.1 Oscillator Figure of Merit 83
2 Note About Harmonic Balance in LC Oscillators 84
3 Amplitude in Differential LC Oscillators 85
3.1 Current-Biased Differential Oscillators 85
3.2 Voltage-Biased Oscillator 88
3.3 Colpitts Oscillators 89
3.4 Complementary Differential Oscillator 94
4 Design of Current-Biased Differential Oscillators 96
5 A Design Example 99
5.1 Design Requirements 99
6 Intuitive Explanation of Phase Noise Sources 104
7 Loading in Current-Biased Oscillators 106
8 Sizing of the Current Source Device 109
9 Noise Filtering in Oscillators 110
9.1 Role of the Current Source 110
9.2 Noise Filtering 112
10 Prototype Oscillator 116
11 Practical Considerations 119
11.1 Power Supply Rejection with Noise Filtering 119
11.2 Device Limitations on Maximum Swing 122
11.3 Extending the Tuning Range 123
11.4 Noise Filtering in Earlier Works 125
11.5 Noise Filtering in Other Oscillators 126
12 Example: Redesign of GSM VCO 129
13 Anatomy of the Figure-of-Merit 132

Chapter 6 Flicker Noise 139
1 Flicker Induced Phase Noise 139
2 FM Due to Modulated Frequency Content 139
2.1 Groszkowski in a van der Pol Oscillator 142
2.2 Groszkowski in the Differential Pair Oscillator 144
2.3 Critical Oscillation 146
2.4 Hard Limiting 149
3 Switch Voltage Noise Modulates Capacitance 151
4 Frequency Modulation by the Current Source 155

Chapter 7 Design for Low Flicker Phase Noise 165
1 Introduction 165
2 Flicker Noise Minimization 166
3 Nulling Flicker Noise 168
4 Wideband Nulling of Flicker Noise Up Conversion 172

Chapter 8 The Role of the Varactor 175
1 Fundamentals 175
2 Types of Varactors 179
3 Varactor Tuning 180
4 Analytical Evaluation of Noise Sensitivity 181
5 AM-to-FM Noise Conversion 186
6 Tuning and Supply Sensitivity 187
7 Measurements and Simulation Results 190
8 Discussion 194
Appendix A 194
Appendix B 197
Appendix C 200

Index 203

Preface

Oscillatory systems exist everywhere, from our planet circulating around the sun with a period of 365.2422 days in an average tropical year; to a pendulum in an antique clock ticking every second; to the vibrations of a quartz crystal in a wrist watch. The study of oscillators was initiated centuries ago in basic mechanics. Some of the very complicated problems of injection locking in coupled oscillators were experimentally verified in the 17th century by Huygens. He used coupled pendulums using elastic threads to move energy from one pendulum to another. Oscillators belong to a class of systems known as *autonomous* systems. As opposed to driven systems, oscillators possess the unique feature that they do not need a time varying input to produce a time varying output. The periodicity and amplitude of the produced oscillation are regulated by the system's energy balance rather than an external input. This unique property makes the study of oscillators both complicated and fascinating.

In the field of electrical circuits, the study of oscillators was pioneered by radio scientists and particularly flourished during World War II. Some ingenious circuit implementations were devised to produce the best oscillators possible. Along with the circuit implementations, came the formal mathematical analysis. One of the earliest models is due to Van Der Pol in the 1920s. Rigorous nonlinear analysis was carried out throughout the 1920's until today.

Despite the long history, most of the literature, until recently, focused on two questions: 'what is the precise amplitude of oscillation?' and 'what is the exact period of oscillation?'

The question of noise behavior was addressed much later. The work of Edson was one of the earliest to discuss the output spectrum of an autonomous oscillator in circuit terms. The work of Leeson in 1964 was perhaps one of the first to address *phase* noise as a distinct class of noise in electronic oscillators and try to predict it using mathematical expressions. His *heuristic model without mathematical proof* is almost universally accepted. However, it entails a circuit specific noise factor that is not known a priori and so is not predictive.

In this work, we attempt to address the topic of oscillator design from a different perspective. By introducing a new paradigm that accurately captures the subtleties of phase noise we try to answer the question: 'why do oscillators behave in a particular way?' and 'what can be done to build an optimum design?' It is also hoped that the paradigm is useful in other areas of circuit design such as frequency synthesis and clock recovery.

In Chapter 1, a general introduction and motivation to the subject is presented. Chapter 2 summarizes the fundamentals of phase noise and timing jitter and discusses earlier works on oscillator's phase noise analysis. Chapter 3 and Chapter 4 analyze the physical mechanisms behind phase noise generation in current-biased and Colpitts oscillators. Chapter 5 discusses design trade-offs and new techniques in LC oscillator design that allows optimal design. Chapter 6 and Chapter 7 discuss a topic that is typically ignored in oscillator design. That is flicker noise in LC oscillators. Finally, Chapter 8 is dedicated to the complete analysis of the role of varactors both in tuning and AM-FM noise conversion.

In some sense, oscillators are the last of obscure analog circuits. The purpose of this book is to put together a sensible theory and optimization methodology. The objective is to lead the reader to understand and efficiently design oscillators using a mechanistic approach that does not entail complicated mathematics yet gives accurate results and design insights.

Acknowledgements

The authors are indebted to numerous people for their help with this book. First, our editor, Ken Kundert with his knowledge, skill, and dedication. Without his tireless efforts, this book would have never been published. We are also grateful to Carl Harris from Kluwer for his patience and encouragement.

We would like to thank our wives, Shahinaz Shahin and Elia Perez, as well as our parents.

The authors would like to thank Linda Jaramillo, Henry and Sarah Rael. We would also like to thank Nikolaus Klemmer for multiple discussions and Wen Suter for encouragement.

Emad Hegazi
Jacob Rael
Asad Abidi

August 12, 2004

1 Basics of LC Oscillators

1 Introduction

Phase noise has been one of the most interesting yet poorly understood topics in circuit design. The challenge of predicting the amount of phase noise in a given circuit has been approached from a variety of angles, from using a "heuristic model without formal proof" [1] to simulation techniques that lead to accurate results [2][3]. Completing this book involved hours of computer simulation, pages of algebra, translation of antiquated notation, and utilizing archeological-type efforts to unearth significant but otherwise forgotten papers.

With regard to this work, three major steps lead to the solution of this problem. The first is Lesson's equation. All works on phase noise must reference Lesson's equation because it is simple, intuitive, and has withstood the test of time [1]. The second innovation was in the development of Cadence, Inc.'s SpectreRF simulation tool [2]. This tool accurately predicts phase noise and served as a test bench to validate all derived equations. Finally, Huang [4] showed it was possible to write out equations for phase noise explicitly. This work takes inspiration and elements from all these works and forms a model that is as intuitive as Leeson, as accurate as SpectreRF, and as rigorous as Huang.

2 The Mathematical Oscillator

The mathematical model of an ideal voltage-controlled oscillator starting at $t = 0$ is described by the following expression:

$$v(t) = A(t)\sin\left(\omega_o t + K_v \int_0^t v_c(u)du\right), \tag{1}$$

where v_c is the control voltage of the oscillator as a function of time. The center frequency of oscillation is ω_o and the instantaneous frequency of oscillation is given by:

$$\omega(t) = \omega_o + K_v v_c(t), \tag{2}$$

where K_v is the oscillator sensitivity and typically given in rad/sec/volt.

As can be seen from (1), the small signal model of an oscillator in frequency/phase domain, with voltage as an input and phase as an output, functions as an integrator. The frequency of oscillation is directly proportional to the control voltage and the oscillator phase is the time-integral of frequency. Being a self-timed system, the oscillator lacks the ability to correct for its own phase. Imagine an oscillator running at some frequency with a constant bias applied to its control voltage line. Any disturbance on the control line will result in instantaneous frequency shift that integrates over the time the disturbance lasts. The resulting phase error will last indefinitely and can never be recovered even though the disturbance lasted for a short amount of time as shown in Figure 1.

FIGURE 1 *Phase jitter accumulation.*

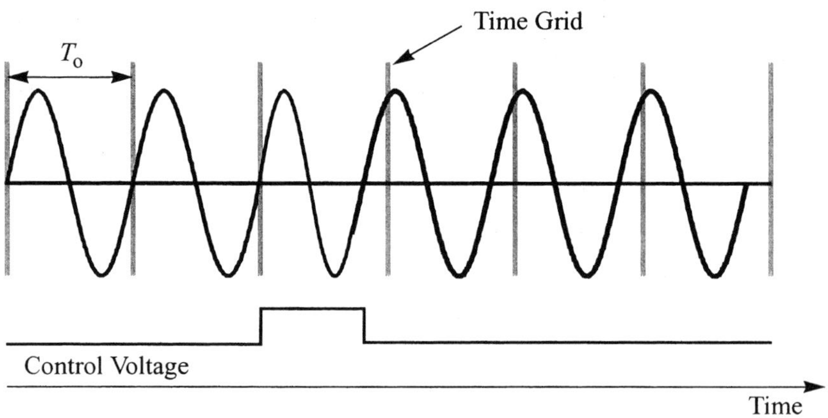

3 Additive White Noise in LC Oscillators

An oscillation is fully characterized by its amplitude and phase. When white noise is added to an oscillation, noise corrupts both the amplitude and phase of oscillation. Assume a noise signal $n(t)$ added to an oscillation $v(t)$. It is customary to model noise as an infinite number of uncorrelated sinusoids separated by 1 Hz each. The sum of the oscillation signal at ω_o and a noise signal at frequency $\omega_o + \omega_n$ is given by:

$$s(t) = A_o \sin(\omega_o t) + a_n \sin((\omega_o + \omega_n)t + \phi_n), \tag{3}$$

where a_n is the noise amplitude and ϕ_n is a random phase. This equation can re-written as

$$\begin{aligned} s(t) = A_o \sin(\omega_o t) &+ \frac{a_n}{2}\sin((\omega_o + \omega_n)t + \phi_n) + \frac{a_n}{2}\sin((\omega_o - \omega_n)t - \phi_n) \\ &+ \frac{a_n}{2}\sin((\omega_o + \omega_n)t + \phi_n) - \frac{a_n}{2}\sin((\omega_o - \omega_n)t - \phi_n). \end{aligned} \tag{4}$$

The first three terms constitute an amplitude modulated carrier with the modulating tones at ω_n. The last two terms together with the carrier, approximate a narrow band phase modulation signal. This means that a single sideband noise component added to the oscillator modulates both the amplitude and phase of the oscillation. The power of amplitude modulation sidebands is equal to the power of phase modulation sidebands.

In Chapter 2, we will rigorously define phase noise. For now, any noise that modulates the phase of oscillation is phase noise. Any noise that modulates the amplitude is considered amplitude noise and is unimportant in most practical cases, except when it later converts to phase or frequency noise.

4 The Linear Oscillator

4.1 Warning

In this section, we develop a *misleading* analysis of oscillators based on linear system theory. Despite looking reasonable, we will show later why it is not accurate or even correct.

4.2 Linear System Theory Applied to Oscillators

Oscillators are fundamentally nonlinear. In fact oscillator's nonlinearity is the reason for their "stable" amplitude. Yet linear models are often used to describe oscillatory behavior. This is acceptable when oscillation start-up conditions are pursued because oscillation at start-up is a small signal. However, the periodically stable frequency can be far different from the small signal "linear" prediction. Furthermore, a linear oscillator model cannot predict the oscillation amplitude. In fact, the assumption of linearity, leads to an undetermined amplitude. This is because in a linear system, if the input doubles, the output doubles. In an oscillator, this leads to an amplitude that is arbitrary [5].

So what is a linear model good for?

1. It can yields a startup condition for oscillation, and
2. it gives a rough estimate of the frequency of oscillation.

FIGURE 2 *Basic LC oscillator.*

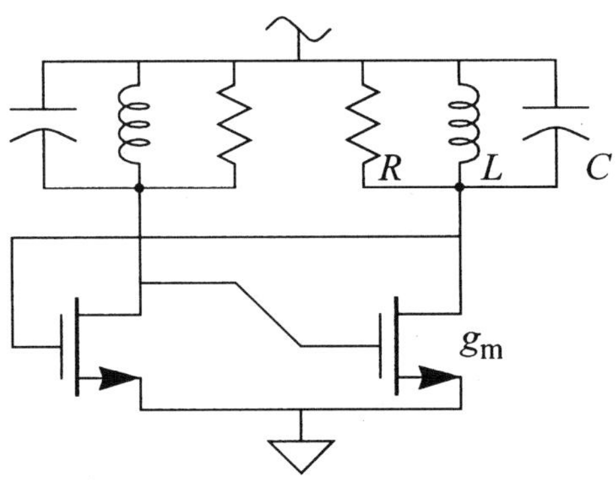

Consider the *LC* oscillator shown in Figure 2. If the oscillator loop is cut at any point, the gain around the loop is given by:

$$G(j\omega) = \left(\frac{-g_m}{\frac{1}{R} + j\omega C - \frac{j}{\omega L}}\right)^2. \tag{5}$$

For a sustained oscillation, *Barkhausen criterion* mandates that the gain around the loop is exactly unity and the phase shift around the loop is precisely 360 degrees. This leads to the following:

$$g_m R = 1, \tag{6}$$

$$\omega_o C = \frac{1}{\omega_o L}. \tag{7}$$

The oscillator shown in Figure 2 can be modeled as a positive feedback system. In Figure 3, the oscillator is constructed using an amplifier and a phase shift network. The amplifier provides no phase shift. The modes of oscillations for this system occur at the natural frequencies of the phase shift network. At these frequencies, the phase shift of this network is a multiple of 360 degrees. If there are multiple frequencies at which Barkhausen criterion is met, then the oscillator can have multiple modes of oscillation. The mode with the highest gain is most likely to prevail but multiple modes of oscillation can coexist.

FIGURE 3 *Feedback amplifier model of the oscillator.*

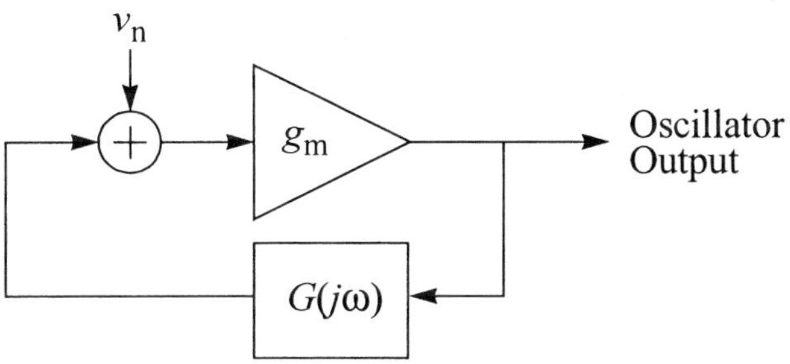

Another way to model an oscillator is a single port model. A lossless LC tank is an oscillator with its frequency of oscillation that can be computed from (7). Loss in the tank damps the oscillation with a time constant equal to $1/RC$. Adding a negative resistance element replenishes any current that flows through the lossy element to sustain oscillation. If the oscillation is to grow then the energy supplied by the negative resistance element must equal the energy lost per cycle. For the oscillator in Figure 2, the two transistors are arranged such that they provide a negative resistance of $-2/g_m$. The differential resistance of the tank is $2R$. Therefore, for a sustained oscillation, the negative resistance should be equal to the positive resistance at all times. This leads back to (6). Redrawn in Figure 4, this circuit is known as the *voltage-biased* oscillator.

FIGURE 4 *Basic LC oscillator (redrawn).*

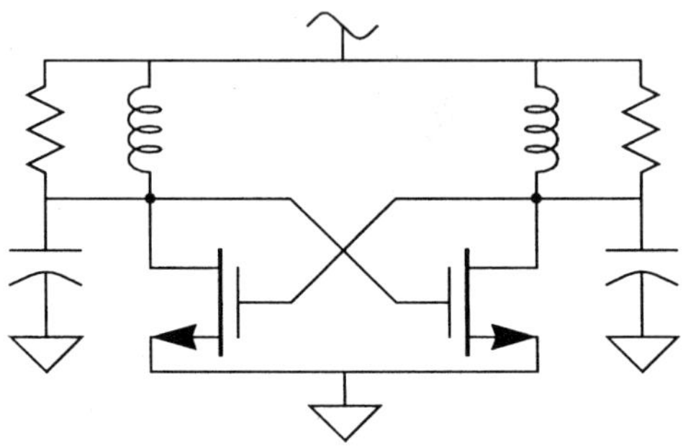

5 Linear Oscillator Noise Analysis

Imagine an oscillator constructed using a parallel LC tank and a transconductor in a positive feedback loop. The frequency of oscillation is of course given by (7). Note that the resonant frequency of the LC tank is the same as the oscillation frequency because both the resistor and the transconductor do not provide any extra phase shift as they carry no reactive current. The reader can readily prove that if the resistive loss is modeled in series with the inductor rather than it parallel, the oscillation frequency will be different from that given by (7).

Now let's consider the noise. Noise can come from two sources in this system: the resistor and the transconductor. Resistor noise is modeled by a white thermal noise current whose density is given by:

$$\hat{i}_n^2(f) = \frac{4kT}{R}. \tag{8}$$

Noise in the transconductor is also modeled as a white noise current whose density is given by:

$$\hat{i}_n^2(f) = 4kTg_m\gamma, \tag{9}$$

where γ is the noise figure of the transconductor element.

The combined noise can be referred to the input of the transconductor as a white noise voltage given by:

$$\hat{v}_n^2(f) = \frac{4kT\gamma}{g_m} + \frac{4kT}{g_m^2 R}. \tag{10}$$

Substituting from (6), the input referred noise voltage at the transconductor input is given by:

$$\hat{v}_n^2(f) = 4kTR(1+\gamma) = 4kTFR, \tag{11}$$

where $F = 1 + \gamma$ is the noise figure of the entire oscillator.

The tank impedance at a frequency $\delta\omega$ away from the resonance frequency can be approximated by

$$Z(\delta\omega) = \frac{R}{1 + j2Q\frac{\delta\omega}{\omega_o}}, \tag{12}$$

where Q is the tank quality factor.

Using basic feedback theory, it is trivial to prove that the closed-loop transfer function from the noise input to the oscillator output is given by:

$$|H(\delta\omega)|^2 = \left(\frac{\omega_o}{2Q\delta\omega}\right)^2. \tag{13}$$

In noise analysis, it is customary to represent noise by a sine wave in a 1 Hz bandwidth. Let's consider a noise component at a frequency $d\omega$ away from the carrier (i.e. oscillation fundamental tone). Noise power at the output of the oscillator can be deduced using (11) and (13),

$$N_{out}(\delta\omega) = 4kTFR\left(\frac{\omega_o}{2Q\delta\omega}\right)^2. \tag{14}$$

As shown earlier, additive noise shows as half amplitude noise and half phase noise. The *noise-to-carrier* ratio is obtained by dividing the output phase noise by the carrier power. We should also consider noise in the lower side band at $\omega_o + \delta\omega$. The single side band noise to carrier ratio is obtained by adding noise power at $+\delta\omega$ and $-\delta\omega$ resulting in the following expression:

$$\mathcal{L}(\delta\omega) = \left(\frac{4kTFR}{P_o}\left(\frac{\omega_o}{2Q\delta\omega}\right)^2\right)\frac{2}{2} = \frac{kTFR}{P_o}\left(\frac{\omega_o}{Q}\right)^2\frac{1}{\delta\omega^2}. \tag{15}$$

This is the renowned Leeson's equation [1]. In the original paper it was given as a *heuristic equation without formal proof.* The preceding analysis is not part of the original paper that was based on measurements and observations. Many other researchers and design engineers derived, over the years, proofs similar to the one we derived here and ended up with one version or another of (15).

What's wrong with the above analysis? First, it is linear and time invariant. Therefore, no frequency translations of noise can occur. This means that low frequency noise, such as flicker noise, cannot create phase noise under the assumptions of this model. The only type of noise that can create phase noise in this model is noise originating around the oscillation frequency. Moreover, it has to have an equivalent amount of amplitude noise because it is in essence, additive noise. In any *LC* oscillator, this is not true. As we will show later, some elements contribute pure phase noise and no amplitude noise. Finally, linear analysis cannot predict the amplitude of oscillation. The amplitude limiting mechanism is fundamentally nonlinear and cannot be captured in the context of a linear time-invariant analysis.

6 How Is This Book Different?

In the following chapters we will show in detail why the derivation in Section 4 is wrong. We will show how to use circuit theory to derive an accurate model for phase noise in electrical oscillators. We will describe what we call a mechanistic model that captures the dominant nonlinearities in an oscillator and provides a closed form expression for phase noise. No fudge factors utilized!

Concepts from nonlinear circuit theory are sometimes utilized yet we tried to keep that to the minimum necessary.

By doing so, this book provides deep insight into the operation of oscillators and provides simple procedures for designing high-purity oscillators. We are answering the seldom tackled questions: '*why* does the oscillator behave that way?' and '*how* is an optimal oscillator designed?'

References

[1] D. B. Leeson, "A simple model of feedback oscillator noise spectrum," *Proceedings of the IEEE*, vol. 54, pp. 329-330, 1966.

[2] K. S. Kundert, "Introduction to RF simulation and its application," *IEEE J. Solid-State Circuits*, pp. 1298-319, 1999. Also available from *www.designers-guide.com/Analysis*.

[3] A. Hajimiri and T. H. Lee, "A general theory of phase noise in electrical oscillators," *IEEE Journal of Solid-State Circuits*, vol. 33, pp. 179-94, 1998.

[4] Q. Huang, "On the exact design of RF oscillators," *Proceedings of the IEEE 1998 Custom Integrated Circuits Conference*, 1998.

[5] Thomas Lee, *Radio Frequency Integrated Circuits*, Cambridge University Press, 1998.

2 Oscillator Purity Fundamentals

1 Introduction

A sinusoidal oscillation is described completely by its amplitude and frequency. The purity of the oscillation pertains typically to only the frequency or period of oscillation. This is because an oscillator is typically used to synchronize events in time. This applies to wireless transceivers as well as digital circuits. The amplitude of oscillation, once above a threshold value, is typically irrelevant as long as it can generate an action (event) on the subsequent circuit block. An oscillation that is perfectly pure has a constant period of repetition. In frequency domain, the perfect repetition translates into a single tone, for a sinusoidal oscillation anyway.

Historically, observing periodicity in time predates the observation of spectral purity for obvious reasons. Ancient Egyptians noted that Sirius rises to its place besides the sun exactly every 365 days. They divided the year into twelve 30-day months and a short 5 day month that was called the "little month". Their calendar is still used today by peasants in the countryside of Egypt, side-by-side with the Gregorian calendar, as it fits perfectly the Egyptian climate and the flooding of the Nile. Their calendar was the basis of the modern calendar we use today. The Babylonians of today's Iraq used a lunar calendar that follows the periodicity of the moon, a cycle of 29 to 30 days. They are the ones who divided the day into smaller units that relate to their base-60 numeral system. Around the 16th century, in the era of great expeditions, there was a great need for accurate clocks for navigation. Determining longitude accurately was not possible without a tool that can tell time with great accuracy [1]. Galileo sketched out the concept of the first pendulum-based clock. It was Harrison who first implemented a clock that can tell time within one second of error per day [1]. Various designs of mechanical movements were implemented over the centuries that followed and are perfected today in Swiss mechanical watches. The 1920s witnessed the first quartz clock that enabled much higher frequency/period stability. Later, this fueled the clock industry in Japan. For more accurate time bases, the atomic clock is used as a calibration source. The search for the most accurate clock is in essence the search for a high long-term stability signal source. The question of noise in signal sources was not of concern until World War II. Only then was the study of noise in the oscillator's phase was born. In essence, phase noise relates to the short-term stability of the oscillator's period, frequency, or phase.

Signal purity measures can be divided into two main categories: deterministic and stochastic. Deterministic *impurity* comes from spurious signals that show in the signal spectrum as delta-Dirac impulses, known as *spurs* at a fixed frequency offset from the main tone. Stochastic *impurity* arises from stochastic variation of signal phase and are manifested in noise skirts around the fundamental frequency. Another way of quantifying stochastic purity is by looking at the signal in time domain where stochastic perturbations are manifested as perturbations in the zero crossings of the sinusoidal waveform. For practical RF receiver design purposes, amplitude perturbations are typically of little concern because mixers are not sensitive to them. In transmitters, however, the situation is somewhat different. Amplitude noise would interfere with neighboring channels just as phase or frequency noise does if it spills out of its allotted bandwidth.

In this chapter we briefly describe the basic concepts of signal purity both in time and frequency domains. The reader who is familiar with these concepts can skip this chapter and advance to chapter 3.

2 Timing Jitter

A pure oscillation repeats in time precisely every T seconds, where T is called the oscillation *period*. In other words, if we set a particular threshold voltage level, the oscillation waveform will cross this threshold in a given direction precisely every T seconds. In the presence of noise, the points in time where the oscillation waveform crosses this threshold are dithered around their ideal noiseless locations. In statistical terms, the oscillation waveform is a random process described by:

$$V(\phi_n, t) = A\sin(\omega_o t + \phi_n), \tag{1}$$

where ϕ_n is a stochastic process that produces a random phase fluctuation.

Consider now different realizations of this random process, i.e. different possible waveforms that fulfill (1). Such realizations are shown in Figure 1. The first waveform is an ideal noiseless oscillation. The other two waveforms are sample realizations for different values of $\phi_n(t)$ that have the same statistical properties. The threshold is held at the $V(t) = 0$. As shown, each of the two realizations rises through the threshold at random time points around the ideal waveform crossing points. The difference in the crossing time between the ideal waveform and the various realizations is a random process denoted by Δ_{ji}. It is assumed that all of the realizations of the random process $V(\phi_n, t)$ start from the same zero initial phase. The subscript j denoted the realization number whereas the subscript i denotes the count of the threshold crossing starting from the initial phase at $i = 0$.

At any particular zero crossing i, the difference in crossing times is a random variable that has some particular mean and variance. Consider the zeroth crossing, i.e. at the initial point; all values of Δ_{j0} are equal to zero because all realizations of $V(\theta, t)$ start at the same initial phase. Therefore, the random variable Δ_{j0} has zero mean and zero variance. In other words, the probability distribution of Δ_{j0} is an impulse,

$$\langle\Delta_{j0}\rangle = 0, \langle\Delta_{j0}^2\rangle = 0, P(\Delta_{j0}) = \delta(\Delta), \tag{2}$$

where δ is the delta-Dirac impulse.

Now comes into play, the most characteristic property of oscillators; they are autonomous circuits. The phase of the oscillator is determined only from within. This means that the ending point of cycle 1 is the starting point of cycle 2 and the phase error incurred in cycle 1 is carried over without correction to cycles 2 and 3, ... indefinitely. This is manifested in the basic mathematical model of oscillators shown in Chapter 1; the oscillator is a phase integrator. This means that if we try to evaluate the statistical properties of Δ at the end of cycle m (Δ_{jm}), we are actually looking at the accumulation (inte-

FIGURE 1 *Different realizations of a random phase jitter process.*

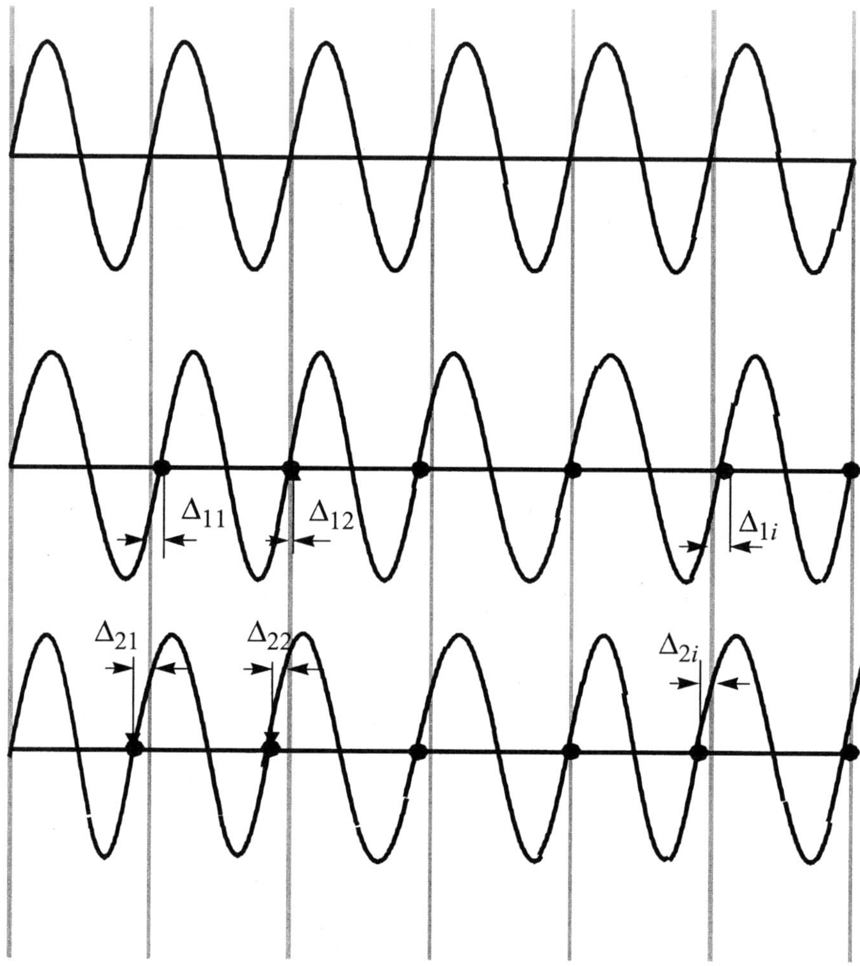

gration) of all Δ_{ji} from $i = 0$ to $i = m$. If we assume for any realization $j = n$, the random fluctuations causing phase jitter are due to white noise (uncorrelated from sample to sample), then the variance of Δ_{ni} grows linearly with time. This was experimentally verified on ring oscillators in [2] but is true for all oscillators. What we have just described is often called absolute jitter [3]. It describes the accumulation of timing jitter with respect to an ideal noiseless source. It was shown that the variance of the absolute jitter Δ as given by McNeil (using his original notation) is:

$$\sigma_{\Delta}^{2} = \kappa^{2} t , \tag{3}$$

where κ is a time domain figure of merit of the oscillator. The higher the oscillator quality factor, Q, the lower the factor κ. The variable t is time or the measurement interval.

The resulting phase jitter from absolute jitter j is given by

$$\theta_{n} = 2\pi \frac{\Delta_{n}}{T} . \tag{4}$$

Note that θ_n is evaluated only at the zero crossings, unlike the phase noise process discussed later, which is defined for all time. Timing jitter measures described here are useful in the world of digital circuits and optical communications clock recovery. Both application domains use square wave like signals and interface to circuits that operate only on the edges of the waveform. Therefore, it is rather common to describe the signal purity in the domains in terms of jitter in picoseconds. In the world of radio-frequency electronics, jitter is almost never used. This is because the signals are sine wave like and phase noise is more relevant as we will see later.

Herzel [4] rewrites (3) as

$$\sigma_{\theta}^{2} = 2Dt , \tag{5}$$

where D is called the *diffusivity*. Ham calls D, the diffusion constant by analogy to the diffusion equation known as Einstein's relation, commonly used in solid-state physics to describe diffusion currents and applicable to any physical diffusion process [5]. This particular treatment is based on the seminal work of Melvin Lax in 1967 in which he analyzed noise in oscillators and showed that the oscillator spectrum (or linewidth) broadens in the presence of noise [6]. We will visit this result in the next subsection. Herzel also draws the same analogy in [7].

As shown by (3) and (5), absolute jitter in a free running oscillator grows indefinitely without bound. Note that we assume that noise in the oscillator is uncorrelated from one cycle to the next, which means that flicker noise is not considered. Flicker noise is correlated from one sample to the next and represents deep system memory. To date, there is no mathematical treatment to describe absolute jitter in a free-running oscillator in the presence of flicker noise. It is also worth mentioning that if the oscillator is locked to a stable source in a phase-locked loop (PLL) absolute jitter does not grow indefinitely. Rather, it grows linearly until the measurement interval is within the loop time constant. In other words, the PLL cannot correct for fast jitter, but it does respond to jitter that is slower than the loop dynamics. This observation is shown in Figure 2 depicting, on a log-log scale, the absolute jitter versus time in a free-running and a locked oscillator.

FIGURE 2 *Absolute jitter in locked and free-running oscillators.*

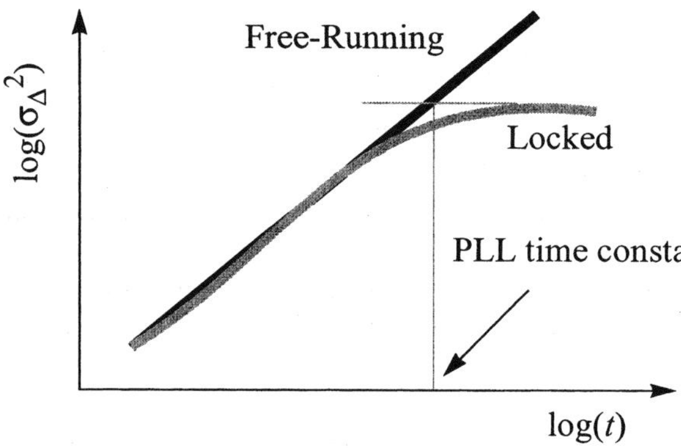

Absolute jitter is not of much use in practice. Being unbounded, little use can be made out of it. A more important jitter measure is called cycle-to-cycle jitter [8] or period jitter [3]. In any oscillation period, this type of jitter is defined as the difference between the true instantaneous period of oscillation and the ideal (or long term average) period of oscillation. The name cycle-to-cycle jitter can be a little misleading and therefore we will use the name period jitter. It can defined as

$$j_n \equiv t_{n+1} - t_n - T, \tag{6}$$

where t_{n+1} and t_n are the end and beginning time points of the nth oscillation cycle [3]. T is the period of noise-free oscillation. If we assume that the oscillation frequency is fixed (pure phase modulation) then T is also the average period of the noisy oscillation. Considering again thermal noise only, the jitter in the nth cycle is uncorrelated with the jitter all other cycles. In this case, it should be obvious that the absolute jitter variance given by (3) is merely the accumulation of σ^2_{jn} over all cycles from $t = 0$ until $t = nT$. Therefore, (3) can be used to predict the RMS value of the period jitter using the substitution $t = T$, i.e. a measurement interval of one oscillation period,

$$\sigma_j^2 = \kappa^2 T. \tag{7}$$

3 Recognizing Phase Noise

Consider a pair of sidebands around an oscillation frequency, as shown in Figure 3. Depending on the relative phase of these sidebands, one can have amplitude modulation or phase modulation. If, for instance, the sidebands sum together such that the sum is co-linear at all times with the carrier phasor, amplitude modulation results. On the other hand, if one of the sidebands is flipped, the sum is always orthogonal to the carrier, in which case phase modulation results. This idea will be used in the subsequent analysis.

FIGURE 3 *Sideband magnitude does not reveal modulation (a). Vector diagram shows that if sidebands sum co-linear to the carrier, AM (b); if orthogonal, PM (c).*

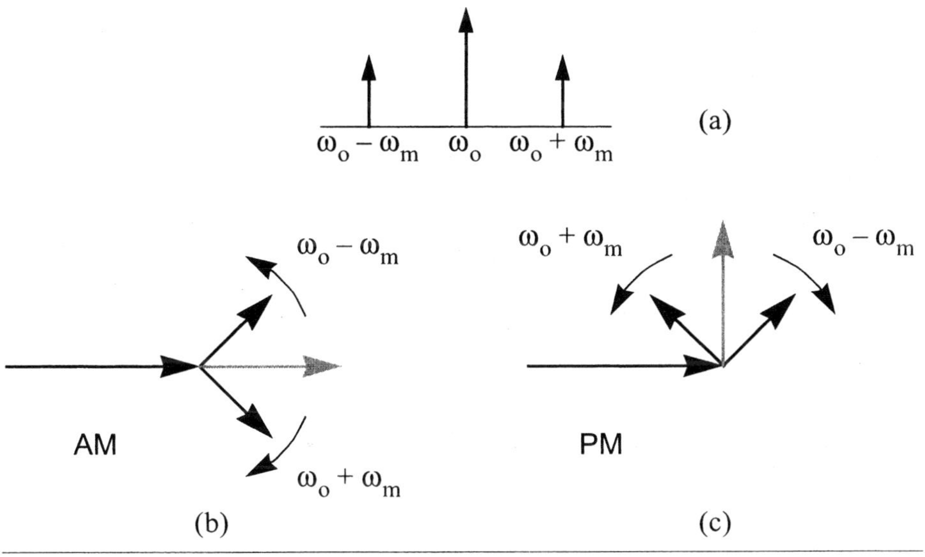

For example, if it is assumed the carrier is a cosine, the various modulation terms are easily derived and given in (8).

$$\begin{aligned} v_{\text{out}}(t) = \; & V_1 \cos(\omega_o t) + \\ & \phi_1(\cos(\omega_l t) - \cos(\omega_u t)) + \alpha_1(\cos(\omega_l t) + \cos(\omega_u t)) + \\ & \phi_2(\sin(\omega_l t) + \sin(\omega_u t)) + \alpha_2(\sin(\omega_l t) - \sin(\omega_u t)), \end{aligned} \tag{8}$$

where the ϕ_1 and ϕ_2 terms are PM sidebands and α_1 and α_2 are AM sidebands.

It is easily seen how each pair contributes either only PM or AM sidebands by drawing a phasor diagram. For the α_1 term, both phasors are parallel to the carrier. Likewise the ϕ_2 pair are orthogonal. The other two cases are easily seen by allowing the phasors to rotate with time until they are either parallel or orthogonal with the carrier.

Working with all of these sine and cosine terms becomes difficult and distracting. A simpler notation can be used to capture the AM and PM components. In (9), the a sideband represents PM and the b sideband represents AM.

$$V_1 = Ae^{j\omega t} + (a_1 e^{j\omega_+ t} - a_1^* e^{j\omega_- t}) + (b_1 e^{j\omega_+ t} + b_1^* e^{j\omega_- t}) \tag{9}$$

Figure 4 illustrates these phasors in the complex domain.

FIGURE 4 *Complex reference phasors indicating PM and AM.*

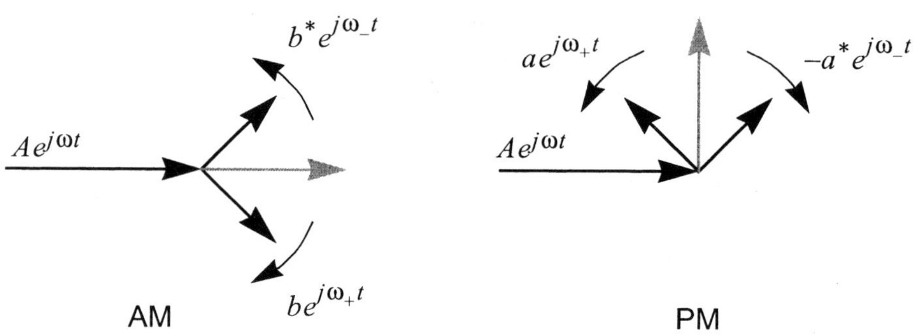

4 Single Sideband Contains AM and PM

A noise spectrum can be approximated by combining a large number of sinusoids whose frequencies are distributed over the range of interest and whose phase is random [16]. In particular, for this analysis, the noise is partitioned into bins of a particular bandwidth, say 1 Hz, and the noise in that bin is approximated by a single sinewave. The frequency of the sinewave is the center frequency of the bin, the amplitude is set such that it has the same power as the noise it represents, and the phase is random. This is a good approximation as long as you observe the noise for a time substantially less $1/\Delta f$, where Δf is the bin bandwidth. The reason is if you observe for longer periods, you can resolve the discrete tones in the approximation and so can distinguish the approximation from the original system.

Consider the case where the noise is combined with a large periodic carrier signal. If the noise is stationary and is added to the carrier, then the phases of each of the sinusoids that make up the approximation of the noise are uncorrelated. In this case, the noise is referred to as "additive noise". Now consider only the noise at a fixed offset frequency $\delta\omega$ from the carrier. As shown in Figure 5, this additive noise can be decomposed into equal amounts of amplitude and phase modulation. Both forms of modulation have components at $\omega_+ = \delta\omega$ and $\omega_- = -\delta\omega$, but in this case the components combine to reinforce each other at ω_+ and cancel each other at ω_-. Thus, additive noise consists of equal amounts of AM and PM noise.

A key observation of this work is that nonlinear circuits respond differently to AM and PM. For example, if an AM input is applied to a limiting amplifier, the output is clipped, removing the AM sidebands, Figure 6. Therefore, the circuit clips voltage or current. However, since it cannot clip along the time axis, PM signals are unaffected. For a PM signal to be unaffected means that the carrier-to-PM sideband ratio of the input is equal to the carrier-to-PM ratio of the output.

5 Phase Noise

Phase noise is characterized using the phase perturbation spectral power in a 1 Hz bandwidth at some offset f_m from the center frequency normalized to the

FIGURE 5 *Single noise component decomposed into AM and PM component.*

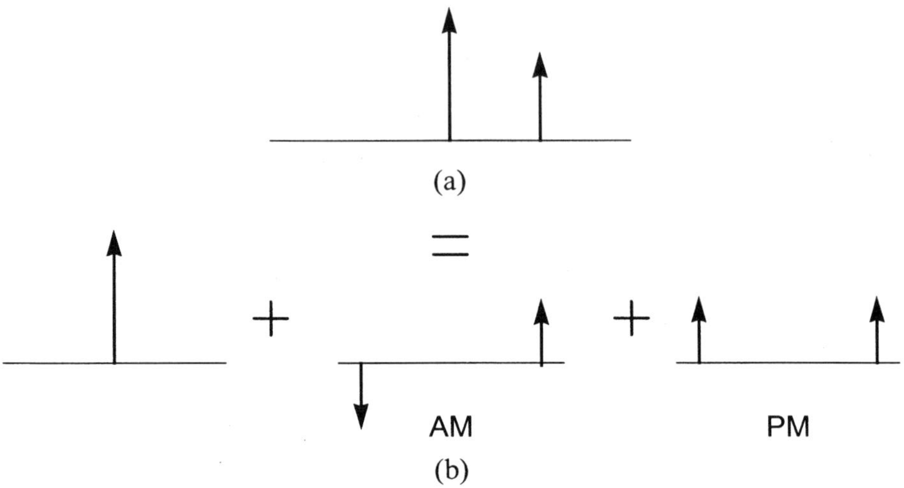

FIGURE 6 *Limiter removes AM sidebands leaving only the carrier.*

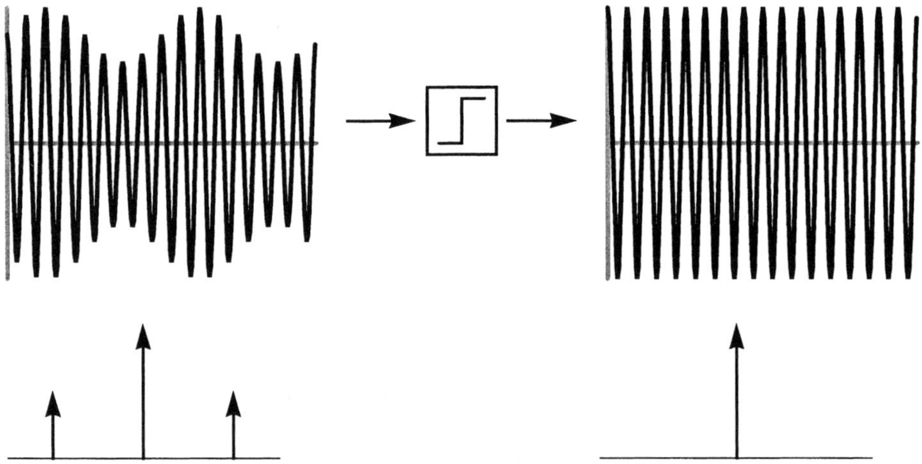

power of the fundamental. If the oscillation waveform is defined by (1) then phase noise is given by:

$$\mathcal{L}(f_m) = \frac{S_v(f_m)}{A^2/2} \tag{10}$$

where $S_v(f_m)$ is the power spectral density of the oscillation signal $v(t)$.

In other words, phase noise is the noise to carrier ratio. Note that this definition does not exclude amplitude perturbations. However, in an oscillator, amplitude noise is naturally rejected by the limiting action inherent in any real implementation.

Graphically, phase noise can be explained as in Figure 7 depicting the output spectrum of an oscillator. When the oscillator is noise-free, its power spectrum is a delta Dirac impulse. An oscillation that is modulated with white noise is known to have the shape of a Lorentzian pulse (the power gain of a first order lowpass filter) [4]. While the total power in the modulated oscillation remains the same for the noisy and the noise-free oscillation, the broadening of the output spectrum is a characteristic of autonomous oscillators. The noise power around the average oscillator frequency is also known as the phase noise skirt. In any non-autonomous (driven) linear or nonlinear circuit, if the input to the circuit is a single tone, the output is an overlay of a single tone and noise. That is to say the output spectrum can be decomposed into a delta-Dirac impulse and a background noise. In oscillators, since the system lacks an external time reference, the output spectrum is not a delta-Dirac impulse except in the noiseless case [6]. Instead, the output spectrum is that of a Lorentzian pulse. This is known as linewidth broadening by analogy with lasers, which are optical oscillators.

To evaluate phase noise numerically using (10), it is first necessary to evaluate the power spectral density of the output voltage. There are a myriad of ways to evaluate this for a generic oscillator. One approach was given by Edson [9,10] and later by Herzel [4] and is summarized below.

The starting point is the oscillation voltage waveform given by (1). The instantaneous frequency noise is given by

$$f_n(t) = \frac{d}{dt}\phi_n \tag{11}$$

Phase noise is described by Wiener process and the power spectral density of phase noise is characterized by its –20 dB/dec slope. Therefore the *frequency*

FIGURE 7 *Phase noise skirt.*

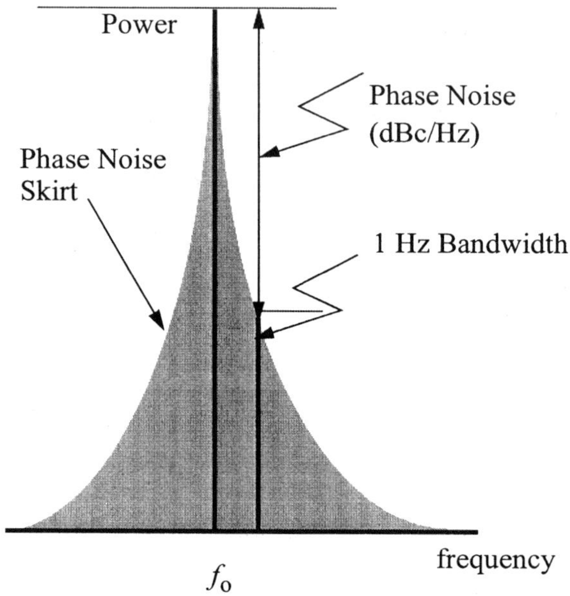

noise spectrum has to be flat [4]. Implicitly, this assumes that the frequency perturbation is a linear function of the noise voltage causing it.

With the assumption that f_n has a white spectrum, the autocorrelation function of f_n is given by

$$R_{f_n}(\tau) = 2D\delta(\tau) \tag{12}$$

Next we look into the autocorrelation function of $v(t)$ as follows

$$R_v(\tau) = E\langle(v(t+\tau))(v(t))\rangle \tag{13}$$

Substituting with (1) into (13), and using basic trigonometry, the following can be readily shown

$$R_v(\tau) = \frac{A^2}{2}([\cos(\omega_o t)E(\cos(\phi(t+\tau)-\phi(t)))] - [\sin(\omega_o t)E(\sin(\phi(t+\tau)-\phi(t)))]) \tag{14}$$

The quantity given by $\phi(t+\tau)-\phi(t)$ is a random variable with a symmetric distribution, which makes the expected value of $\sin(\phi(t+\tau)-\phi(t))$ equal to zero. With this in mind, (14) reduces to:

$$R_v(\tau) = \frac{A^2}{2}([\cos(\omega_o t)E(\cos(\phi(t+\tau)-\phi(t)))] \tag{15}$$

Taking the Fourier transform of (15), we arrive at the power spectral density of $v(t)$:

$$S_v(\omega) = A^2\frac{D}{(\omega-\omega_0)^2+D^2} \tag{16}$$

This shows how the power spectrum of an oscillator takes the shape of a Lorentzian pulse if only white noise is considered. It is clear that the spectrum at zero frequency does not explode to infinity as could happen in a mixer driven by a perfect sinewave.

Substituting in (10), oscillator's phase noise is given by:

$$\mathcal{L}(\omega-\omega_o) = \frac{2D}{(\omega-\omega_0)^2+D^2} \tag{17}$$

For a reasonable quality oscillator, the diffusivity D is a small quantity, meaning that phase diffusion process is slow. Intuitively, this means that the initially synchronized oscillators discussed earlier will take a long time to lose coherence. For frequency offsets $\Delta\omega = \omega-\omega_0 \gg D$, phase noise is given by:

$$\mathcal{L}(\Delta\omega) \approx \frac{2D}{\delta\omega^2} \tag{18}$$

This explains the statement that phase noise at white-noise-dominated offset has a slope of –20 dB/decade. Note again that colored noise is not considered here at all.

It should be clear that D establishes the relationship between phase noise and timing jitter in an oscillator. It should also be noted that we assume that the relationship between the continuous phase modulation ϕ and the discrete-time phase error θ is given by [3]:

$$S_\theta(\omega) = S_\phi(\omega) \tag{19}$$

To find an expression for D, first re-write (7) as follows:

$$\sigma_\theta^2 = 2DT \tag{20}$$

For the nth oscillation period, the discrete phase error θ_n is given by:

$$\theta_n = \omega_o(T_n - 2\pi) = \frac{2\pi}{T}(T_n - T) \tag{21}$$

Accordingly, the period jitter is given by:

$$\sigma_j^2 = \left(\frac{T}{2\pi}\right)^2 \sigma_\theta^2 \tag{22}$$

Finally, we arrive at an expression for D using (20) and (22)

$$\sigma_j^2 = D\frac{T^3}{8\pi^2} \tag{23}$$

Substituting from (23) into (17) or (18) establishes the relationship between jitter and phase noise for an oscillator with only white noise sources. Without much effort, it is trivial to show that

$$\sigma_j = \sqrt{\frac{\mathcal{L}(\Delta\omega)\Delta\omega^2}{\omega_o^3}} \tag{24}$$

6 Oscillator Phase Noise Models: Post-Leeson

The past few years have witnessed an explosion in oscillator phase noise research. Various models for the phenomena are built and evaluated. In addition, multiple simulators are available in the mass market to accurately predict the phase noise performance of the oscillator. The models available are very rigorous. The intuition and design guidelines are not always clear. In this section, we discuss what we consider the most important models in recent years and show briefly the key idea of the model we build and advocate in this book. We will limit the discussion to key ideas without delving into the mathematical details. We will show briefly the strengths and weaknesses of each model.

6.1 Hajimiri's Model

The model was originally presented in [8] as a general theory for phase noise in electrical oscillators. It averts the nonlinear analysis by treating the oscillator as a linear but time-varying system. The model further assumes that noise in an oscillator is a cyclostationary random process. Cyclostationarity means that the first and second-order statistics of the random process are periodic with a period *T*. Where *T* in an oscillator is the period of oscillation. Therefore, the oscillator is called a T-periodic system in that the oscillation waveform repeats itself every *T* seconds. The model builds on the following assumptions:

1. white noise can be treated as uncorrelated random samples in time (impulses)
2. the response of the oscillator to a noise sample depends on the time that sample occurs with respect to the oscillation waveform such that:
   a. noise that occurs at the peak of oscillation can only create amplitude noise.
   b. noise that occurs at the zero-crossings of oscillation can only create phase noise.

Based on these assumptions, Hajimiri develops what is called an Impulse Sensitivity Function (ISF). It measures the sensitivity of the phase of the oscillator to a small perturbation current injected at a particular moment in time. The ISF has the same period *T* of the oscillator itself because of the cyclostationarity assumption. A simulation procedure was also presented in [8], by which means, the ISF can be constructed. Using a SPICE-like transient simulator, the ISF can be evaluated from each noise source in the oscillator to the output. An impulse of current, representing noise in a transistor channel or a resistor etc., is injected into the oscillator at one instant in time. The effect on the oscillator phase is evaluated after multiple cycles when the oscillator is back to its normal limit cycle. The position of the impulse is shifted with respect to the oscillation waveform and simulation is re-run to evaluate the ISF from a particular noise source at a different time point. Using a fairly tedious simulation procedure, the ISF of each noise source is constructed.

To graphically explain the model, consider the waveforms of an oscillation with an injected noise current into the tank. Injecting an impulse current into the tank while the waveform is at the peak, changes only the amplitude of oscillation as shown in Figure 8a. This change in amplitude is rejected by the oscillator due to the presence of the intrinsic nonlinearity of the oscillator and the oscillator manages to restore its original amplitude after a few cycles[†]. Alternatively, if the noise is injected at the zero crossing, only the phase of the oscillation changes as shown in Figure 8b. Phase disturbance does not fade away because the oscillator has no particular phase preference. The accumulated phase jitter is given by the integral:

$$\phi(t) = \frac{1}{q_{\max}} \int_{-\infty}^{t} \Gamma(\omega_o \tau) i_n(\tau) d\tau \tag{25}$$

where q_{max} is the maximum charge on the tank capacitance and τ is the time instant at which noise was injected. Γ is the ISF from the noise current source, $i_n(\tau)$, to the oscillator's phase.

The model assumes linear operation but accounts for time variance. This allows the extension of the model to multiple noise sources once correlation between these sources is accounted for. To account for the time variance of noise in various circuit elements, the noise current is decomposed into a white noise part and a time scaling function as follows:

$$i_n(\tau) = i_{no}(\tau)\alpha(\omega_o t) \tag{26}$$

With the cyclostationarity assumption, α is a T-periodic function.

It is always instructive to consider the oscillator in the state-space plane. An LC oscillator is a second order system, i.e. it possesses two states: capacitor voltage and inductor current. Plotting the inductor current versus the capacitor voltage shows a closed curve (in signal processing terminology it is called a limit cycle). As the voltage across the capacitor changes, so does the inductor current such that the locus repeats every T seconds. However, the oscillator does not necessarily traverse the locus at constant rotational speed. The forgo-

† In an ideal lossless LC tank, the amplitude of oscillation never returns to its original value after a noise impulse. This is because there is no dissipative loss to change the total energy of the tank nor there is nonlinearity to fix the amplitude of oscillation.

FIGURE 8 *Impulse response of an oscillator using Hajimiri's model.*

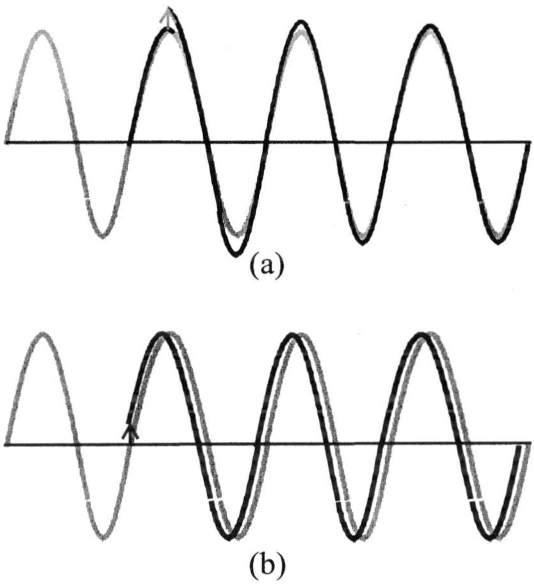

ing assumptions of the model can be explained in the state-space plane shown in Figure 9. Any perturbation can be decomposed into a tangential component to the state-space locus and one perpendicular to it. According to this linear time-varying model, the tangential component disturbs only the oscillator's phase whereas the perpendicular component is pure amplitude noise. Later we will discuss another form of perturbation decomposition.

The model gave some insight into the operation of oscillators. One of its advantages is that it is applicable to any class of oscillators, not only LC or ring-based ones. It provides a simulation procedure that shows reasonable accuracy using a SPICE-like transient simulator and was published at a time where simulators like SpectreRF and EldoRF where not very common. However, it does not formally address flicker noise at all. The treatment of flicker noise in the original paper [8] is performed in the frequency domain rather than the time domain without clear analysis or conclusive answers. The model found wide acceptance at least for its thermal noise treatment. Later, it was deemed accurate only for cases where injected noise is stationary. As a result, the model cannot be used for quadrature oscillators for example [11]. As we

FIGURE 9 *Orthogonal decomposition of perturbation.*

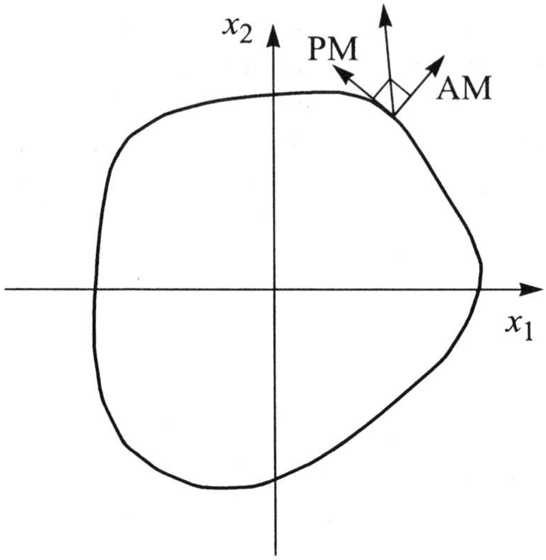

will see in the discussion of Demir's model, the assumption of cyclostationarity leads to an output spectrum that rises to infinity at the oscillation frequency [12]. Therefore, the output spectrum under Hajimiri's model does not resemble a Lorentzian pulse. This particular inaccuracy does not impact the overall accuracy of the model because for a reasonable quality oscillator, the Lorentzian pulse shape dies out quickly as given by (18).

In essence, the model was developed as a simulation methodology. The circuit intuition is gleaned through extensive simulations rather than analytical means. Therefore, there is no systematic way to understand the physics of why a particular topology behaves in a particular way, or what happens if the device width is changed. Despite following efforts to draw more insights, the intuition remains based on extensive simulations and clever observation rather than circuit analysis or physical insight [5].

6.2 Demir's Model

Unlike Hajimiri's model, this work by Demir does not assume cyclostationarity. In fact, the model shows that the oscillator output voltage is a stationary random process. This is justified by the fact that the oscillator has no external time base to set a cyclostationarity period. The model adopts a nonlinear perturbation analysis whereby the perturbation is not decomposed into phase and amplitude noise components. Instead, the perturbation is decomposed into a phase deviation component and an additive component that Demir calls orbital deviation. To illustrate the basic concepts, let's resort to the state-space plane. Consider the oscillator without noise with a limit cycle traversing a certain orbit as shown in Figure 10. When noise sources are considered, the oscillator does not follow the original orbit but can have any orbit within a particular band of orbits. The model encompasses the decomposition of the perturbation into two components:

FIGURE 10 *Kartner's decomposition.*

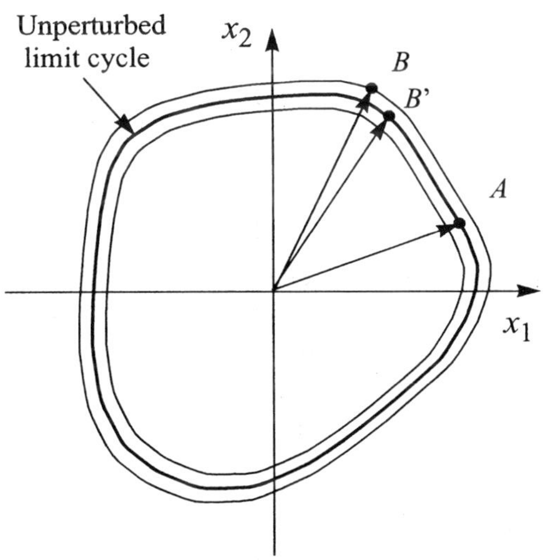

1. Phase deviation component responsible for shifting the phase of the oscillator.

2. Orbital deviation, by which means, the oscillator's limit cycle is momentarily perturbed.

This decomposition was first proposed by Kartner [13]. Demir distinguishes between the two components of perturbation in the following manner: the orbital deviation does not accumulate and its impact on the oscillator's limit cycle dies to zero if the perturbation is removed. Phase deviation, on the other hand, accumulates and if the perturbation source is removed the phase error produced during the time of perturbation remains indefinitely. Consider the state-space plot in Figure 10. At a particular point in time, the oscillator was at point A on the unperturbed locus. If noise is included, the accumulation of phase error can result in the oscillator's state being at point B instead. Clearly, point B is not close to point A as the phase error can grow without bound. Point B however, can be projected on another point B' that falls on the unperturbed limit cycle as well as a small component for B' to B. The translation of the oscillator's phase from A to B' is caused by the phase deviation component. The component from B' to B is the orbital deviation. Using a relatively complicated mathematical description, entailing stochastic differential equations, Demir manages to find a proper and formal description of the output spectrum of the VCO in the presence of thermal noise. Of most significance, is the proof that the oscillator output is not cyclostationary but rather a stationary process. The phase of the oscillator is non stationary, as would be expected of an integrated noise. Demir's argument for the physical origin of stationarity in the output voltage of the oscillator is also interesting. He recognizes that the oscillator's period is disturbed by noise, which means that it cannot set a proper cycle for a cyclostationary random process. Using the original terminology, the oscillator is represented by a group of equations in the form:

$$\frac{\partial}{\partial t}x(t) = f(x(t)) \tag{27}$$

where $x(t)$ is the oscillator's output voltage.

When the oscillator is perturbed by a small perturbation $b(t)$, the output voltage takes the form $x(t+\theta(t))+y(t)$, where $y(t)$ is the orbital deviation.

The phase noise resulting from the voltage perturbation $b(t)$ can be obtained by solving the following equation:

$$\frac{\partial\theta}{\partial t} = v(t+\theta(t))B(t+\theta(t))(b(t)) \tag{28}$$

where v is called the perturbation projection vector and plays a similar role to the ISF in Hajimiri's analysis and B captures the response of the system of equations in (27) to the perturbation b.

The proposed solution is acquired using a nonlinear two-step perturbation analysis. Using the perturbation proposed by Kartner, the equation system given by (27) is first solved by assuming no orbital deviation to yield a solution that has only the accumulated phase error. Later, the resulting solution is re-solved including a small perturbation using linear perturbation.

As shown by other researchers, Demir shows that the spectrum of an oscillator is a Lorentzian pulse in the presence of white noise. Demir further extended his modeling to include colored noise source, which Hajimiri's model does not cover. He first redefines the flicker noise spectrum as that which causes 0 dBc at zero frequency in an oscillator and solves a system of stochastic differential equations to find the output spectrum of an oscillator in the presence of colored noise [14].

In a recent publication, a comparison between Demir's work on white noise and that of Hajimiri was drawn [11]. It shows that the two models are identical in the case of stationary noise. It also shows that in the case of quadrature oscillators Hajimiri's equations collapse whereas those of Demir still yield the correct result. Again, Demir's work is valid for all classes of oscillators. No independent verification of Demir's flicker noise work is available to date.

Demir's model is perhaps the most generic and most accurate treatment of noise in oscillators. It is mathematically involved and perfectly suited for a simulator-type application. It does not help circuit designers know how to design the best oscillator or how to improve on existing topologies by showing the mechanisms of phase noise generation. There is almost no intuition into circuits that results from Demir's model but it is definitely a remarkable addition to the simulation of phase noise in oscillators.

6.3 A Mechanistic Physical Model for LC Oscillators

6.3.1 Yet Another Model!

The work presented here can be what is expected from a circuit designer analyzing oscillators. The basic assumptions are closer to those of Hajimiri. We assume a *T*-periodic system disturbed by cyclostationary noise. This cyclostationarity does not appear explicitly in the model equations but it is nevertheless implicit. The model assumes a steady-state solution to the circuit[†], by which means, a large signal sinusoid exists across the tank and random noise sources from circuit elements disturb the amplitude and phase of the oscillator. Unlike Demir and Hajimiri, the work presented here is focused on LC oscillators. Even though the circuit analysis can perhaps be extended to other classes of oscillators, this book does not cover any other than LC oscillators. In essence, the model is built by circuit designers and is intended to be used also by circuit designers.

The model is valid, nevertheless, for both single phase and quadrature oscillators as we will show in the following chapters. Our model handles flicker noise in a systematic way, unlike the linear time-invariant model presented earlier. The model results in closed form expressions rather than a numerical procedure. Therefore, it is best suited for circuit designers who are interested in designing real-life circuits.

6.3.2 Basics of the Approach

Noise from any source in the oscillatory system can be represented by a number of sinusoids evenly distributed every 1 Hz. These are called conformal signals [16]. If the oscillator is disturbed by a small tone at a particular offset with any random phase, the FFT of the oscillator output can be tell what happens to both the amplitude and phase of the injected tone. Due to the large signal operation of the circuit, the gain seen by the injected noise sinusoid can possibly entail frequency translation [15]. Therefore, we call this gain the translation gain. Determining translation gain is possible using transient simulation and an FFT step. It can be also determined using a periodic steady state simulator like SpectreRF. Translation gain has to take noise folding into account. Noise folding occurs because thermal noise has an infinite bandwidth. As a result, noise is under-sampled by both the fundamental tone and the harmonics of the oscillation. Aliases of the under-sampled noise fold to near the fundamental frequency. As shown in Chapter 1 and also in [16], a single-sideband injected into the tank can be decomposed into four components of correlated phases such that it represents both amplitude and phase modulation. The phase information of the FFT can easily show AM and PM sidebands to any degree of accuracy required.

The work presented however, does not resort to simulations to determine the translation gains from each noise source to the output. Instead, we resort to circuit analysis to determine the AM and PM components.

† This is also the underlying assumption in a simulator like SpectreRF

We begin by assuming a steady-state solution, in which there exists the oscillator tone and four sideband modulation tones representing AM and PM. Of course we assume one single source of noise in the oscillator at a time. Each sideband is scaled by an unknown coefficient that needs to be determined using circuit equations in steady state. Once the coefficients are determined, the type of noise (AM or PM) resulting from that noise source is known. The analysis is done for all noise sources. The circuit equations used take into consideration the physical mechanism behind any frequency translations or scaling. More details on the technique will be shown in the following chapter.

Dealing with noise sources in frequency domain rather than time domain has its advantages. The most important one is that it facilitates the analysis of colored noise sources. For example, flicker noise at any frequency is uncorrelated with flicker noise at any other frequency. This greatly simplifies the analysis. However, as we will show, normal frequency translations of flicker noise can only result in amplitude noise. We will also show that flicker noise can create phase noise in the oscillator by means of frequency modulation processes.

The collective analysis presented in the following chapters is what we call a physically-based mechanistic model. It is physical because it relates to actual operation of the circuit elements. It is mechanistic as it describes the behavior of the oscillator using a group of mechanisms rather than a unified numerical methodology. This allows the full optimization of the oscillator by understanding the mechanisms in action that are responsible for phase noise generation. The model is very intuitive while resulting in a reasonable accuracy, within 1 dB of what SpectreRF predicts.

References

[1] David D. Allan, Neil Ashby, and Clifford C. Hodge, "The science of timekeeping," Hewlett-Packard, Application Note 1289, 1997.

[2] John A. McNeil, "Jitter in ring oscillators," *IEEE J. of Solid-State Circuits,* vol. 32, no. 6, June 1997.

[3] David C. Lee, "Analysis of jitter in phase-locked loops," *IEEE Trans. on Circuits and Systems-II Analog and Digital Signal Processing*, vol. 49, no. 11, Nov. 2002.

[4] F. Herzel and B. Razavi, "A study of oscillator jitter due to supply and substrate noise," IEEE *Trans. on Circuits and Systems-II: Analog and Digital Signal Processing*, vol. 46, no. 1, pp. 56-62, Jan. 1999.

[5] D. Ham, and A. Hajimiri, "Virtual damping and Einstein relation in oscillators", *IEEE J. of Solid-State Circuits,* vol. 38, no. 3, pp. 407-418, Mar. 2003.

[6] M. Lax, "Classical noise V. Noise in self-sustained oscillators," *Phys. Rev.*, CAS-160, no. 2, pp. 290-307, Aug. 1967.

[7] F. Herzel, "An analytical model for the power spectral density of a voltage-controlled oscillator and its analogy to the laser linewidth theory," IEEE *Trans. on CAS-I: Fund. Theory and Tech.*, vol. 45, no. 9, pp. 904-908, Sep. 1999.

[8] A. Hajimiri, and T. H. Lee, "A general theory of phase noise in electrical oscillators," *IEEE J. of Solid-State Circuits,* vol. 33, no. 2, pp. 179-194, Feb. 1997.

[9] W. A. Edson, "Noise in oscillators," *Proc. IRE,* pp.1454-1466, Aug. 1960.

[10] Ali Hajimiri, and Thomas H. Lee, *The Design of Low Noise Oscillators*, Kluwer Academic Publishers, 1999.

[11] P. Vanassche, G. Gielen, and W. Sansen, "On the difference between two widely publicized methods for analyzing oscillator phase behavior," *in Proc. of Int. Conf. on Computer-Aided Design, ICCAD*, 2002.

[12] A. Demir, A. Mehrorta, and J. Roychowdhury, "Phase noise in oscillators: a unifying theory and numerical methods for characterization," IEEE Trans. on Circuits and Systems-I Fundamental Theory and Applications, vol. 47, no. 5, pp. 655-674, May 2000.

[13] F. K. Kartner, "Analysis of white and f^{-1} noise in oscillators," *Int. J. Circuit Theory Appl.*, vol. 18, pp. 489-519, 1990.

[14] A. Demir, "Phase noise and timing jitter in oscillators with colored-noise sources," *IEEE Trans. Circuits and Systems-I: Fundamental Theory Appl.* vol. 48, no. 12, pp. 1782-1791, Dec. 2002.

[15] Joel Phillips and Ken Kundert, "Noise in mixers, oscillators, samplers, and logic: an introduction to cyclostationary noise," *www.designers-guide.com/Theory.*

[16] W. A. Robins, *Phase Noise in Signal Sources*, *IEE Press*, 1982.

3 Current Biased Oscillator

The well-known tail-current biased differential LC oscillator is analyzed in this chapter. This oscillator is very popular and has been analyzed by many people [1,2]. Because it is a truly-differential circuit, the analysis here is greatly simplified.

1 Steady-State Operation

The tail-current biased differential LC oscillator is shown in Figure 1a. It consists of a differential pair, which commutates a tail current to provide an effective negative resistance across the resonator. The resonator is drawn as shown because differential inductors provide a higher Q and consume a smaller area than single-ended inductors. The I-V characteristic of the cross-coupled differential pair characteristic is shown in a piecewise linear fashion in Figure 1b. In the linear region, the I-V characteristic has a negative slope that corresponds to the negative resistance of the device. The curve then flattens out and the effective admittance is zero. In this region, the circuit does not provide any negative resistance nor does it provide a loss that would load the resonator. The curve does not have any hysteresis or "loops" and so indicates that the circuit is memoryless.

In steady-state, the oscillation switches this differential pair to produce a differential square-wave current at the output that excites the resonator, making up for losses in the resistor, Figure 2a. The square wave has a frequency spectrum consisting of a fundamental, with amplitude $2I_0/\pi$, and odd harmonics. The resonator filters out the higher harmonics and the fundamental flows through the resistors, with a value R, to determine the differential steady-state amplitude of oscillation. The amplitude is given in (1) and is shown graphically in Figure 2b, [3]. Note that the current in each of the switching transistors switches between 0 and I_0 with a bias value of $I_0/2$. The *signal* current

FIGURE 1 *(a) A cross-coupled differential oscillator and (b) the differential IV characteristic of the cross-coupled differential pair.*

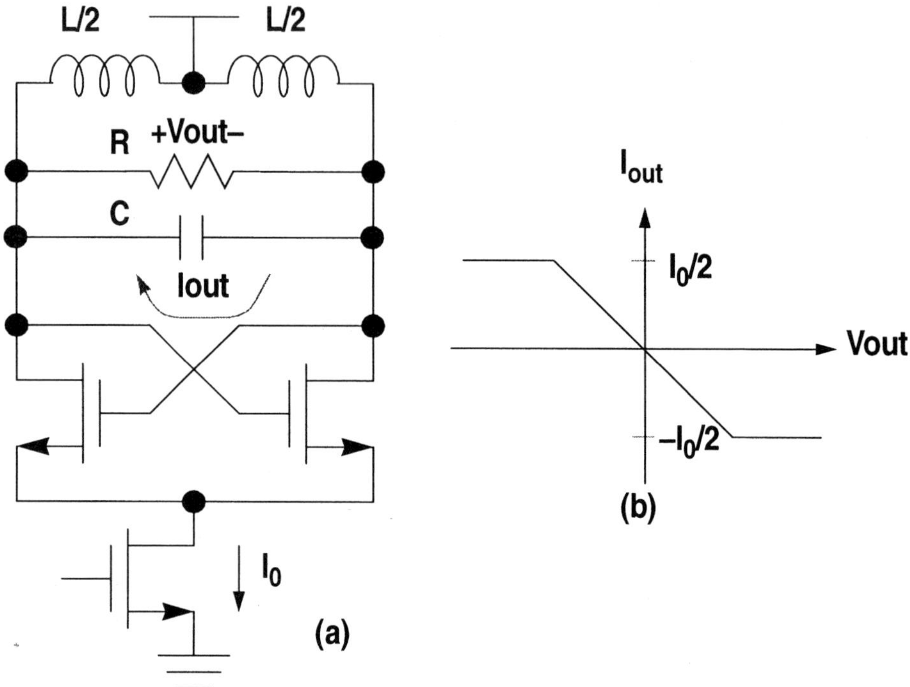

into the differential tank resistance is thus switching between $I_0/2$ and $-I_0/2$. Some references use a different notation where the output current is taken as the difference between the currents in each side of the oscillator. In that case, the switching waveform moves between $-I_0$ and I_0 but flows into $R/2$ to build the amplitude.

$$V_{out} = \frac{2}{\pi} R I_0 \tag{1}$$

The frequency of oscillation is determined by the resonator and occurs when the energy in the inductor is balanced by the energy in the capacitor. If the current flowing into the resonator consists of only a sinewave, the balance occurs at the resonance frequency of the tank.

$$\omega_0 = \frac{1}{\sqrt{LC}}. \tag{2}$$

FIGURE 2 *(a) Cross-coupled differential pair output current in time and frequency domain. (b) Output differential voltage in time and frequency domain.*

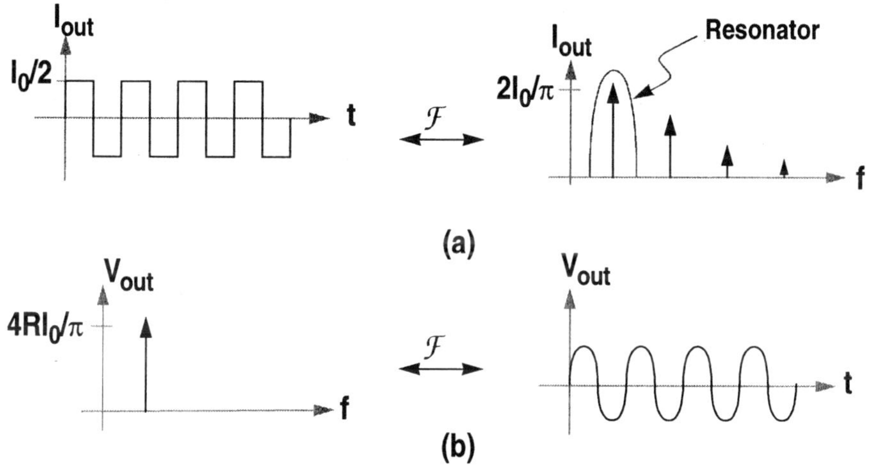

As the tail current is increased, the amplitude of oscillation rises in proportion to the tail current, as shown in Figure 3. However when the swing is so large that the current source is driven into the triode region, the oscillation clips at a level of $2V_{dd}$. At this point, the current source is supplying as much current as the oscillator can support. Here, the current source can be replaced with a short to ground [4]. The linear region of operation is called *current limited* and the clipped region is called *supply limited* [1].

2 Linear Analysis of Differential Oscillator

Phase noise in an LC oscillator is typically modeled by the ratio defined by Leeson and classically given in (3), [5]. Here, P_s is the power of the signal and F is a circuit specific constant.

FIGURE 3 *Initially, as the current is increased, the amplitude is current limited and is proportional to the current. Once the current source is driven into the linear region, the amplitude becomes supply limited.*

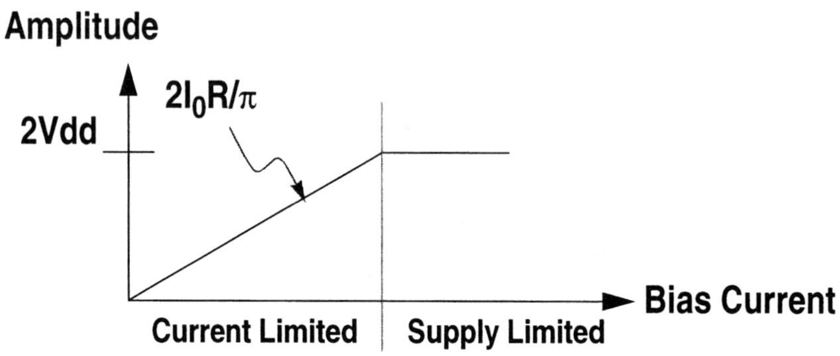

$$\mathcal{L}(\omega_m) = \frac{4FkT}{V_1^2}\left(\frac{\omega_o}{2Q\omega_m}\right)^2 \qquad (3)$$

Leeson's ratio is re-written in (4). Here the noise, kT/C, is filtered by the resonator and normalized by the power of the signal [4].

$$\mathcal{L}(\omega_m) = F\frac{1}{V_1^2}\frac{kT}{C}\frac{\omega_o}{Q}\frac{1}{\omega_m^2} \qquad (4)$$

This ratio is scaled by a circuit-specific noise factor F. This noise factor is similar to other noise factors used in RF circuits like LNAs, except instead of normalizing the total circuit noise to that of 50Ω, it is normalized to the noise of the resonator loss [4].

Since all resonators are lossy, an active element is required to sustain oscillation. Consider Figure 4a, where an ideal negative resistance is used to compensate for the resonator loss. The noise of the negative resistance is equal to the resonator loss but is not correlated with it. This results in a noise factor of 2. Nonlinear active circuits are used to implement the negative resistance, as shown in Figure 4b. Assuming the noise factor of the non-linear circuit is γ, the noise factor of the whole oscillator becomes $1 + \gamma$.

FIGURE 4 *Ideal oscillators with noise figure.*

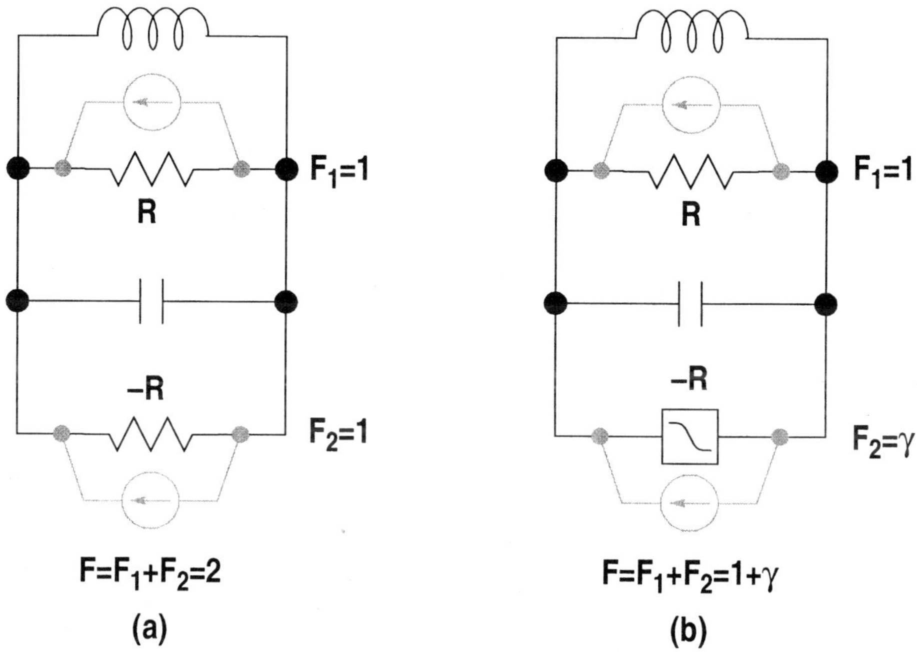

3 Thermally Induced Phase Noise

At high offset frequencies, the major source of phase noise is due to thermal noise. Many techniques have been developed to analyze this process but none provide simple methods that enable simple hand analysis of phase noise in oscillator circuits. Here, the tools to analyze the time-varying nature of phase noise are oscillators are developed.

A little-known property of nonlinear circuits greatly simplifies the analysis of phase noise in oscillators. It can be proven that if a narrow-band, phase modulated signal is applied to a zero-memory non-linear circuit and it is followed by a narrowband filter centered at the carrier, the carrier-to-sidebands of the phase modulation at the input is preserved at the output, Figure 5.

FIGURE 5 *Any memory-less circuit followed by a bandpass filter preserves the PM input carrier-to-sideband ratio at the output. The filter removes any harmonics and sidebands generated by the nonlinear circuit.*

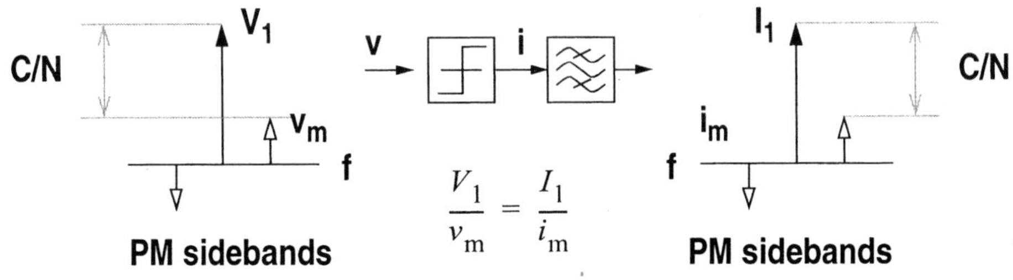

This property is hinted at by Rice [6] and by Belinkiy [7]. Davenport's classic work on limiters [8] compares SNR_{in} to SNR_{out} for white noise but not phase modulated sidebands. Finally, Malyshev has shown that in any non-linear circuit without any reactive elements, no AM to PM conversion occurs (sic. AM-FM conversion) [9].

While this property was originally proved for the differential oscillator using previous work on noise in mixers [11], a more general proof that is easier to apply to a wider class of nonlinear circuits is given here. After reviewing the analysis of various aspects of phase noise from AM to PM in the varactors [12] to frequency modulation in the current source, it was found that the derivation by Samori was the most consistent with the ideas presented here [3]. The following derivation is based on his paper.

Assume the input of a nonlinear transconductor consists of a carrier, V_c, and a lower sideband, V_{lsb}, located ω_m below the carrier.

$$V_m(t) = V_1\cos(\omega_o t) + V_m\cos((\omega_o - \omega_m)t + \phi) \quad (5)$$

If the characteristic of the transconductor is $I_{out}(t) = I(V_{out}(t))$ and the amplitude of V_1 is much larger than V_m, then a Taylor series can be used to approximate the response of the nonlinear circuit.

$$I(V_{out}(t)) \approx I(V_c(t)) + \frac{dI(V_c(t))}{dV}V_{lsb}(t) \quad (6)$$

The derivative dI/dV is the transconductance of the nonlinear device and is shown in gray in Figure 6a.

$$g(V_c(t)) = \frac{dI(V_c(t))}{dV} \tag{7}$$

For the case of a differential pair or a limiting amplifier, the transconductance is an even function with a nonzero value appearing at the zero crossings of the output voltage. This is clearly seen when the transconductance is plotted as a function of time, Figure 6b. The resulting Fourier Series will only contain terms located at even multiples of the carrier frequency,

$$g_c(V_c(t)) = g_0 + \sum_{n=1}^{\infty} g_{2nI}\cos(2n\omega_o t) + g_{2nQ}\sin(2n\omega_o t). \tag{8}$$

FIGURE 6 *(a) Characteristic of a differential pair with transconductance. (b) Transconductance turns on at twice the carrier frequency and is an even function.*

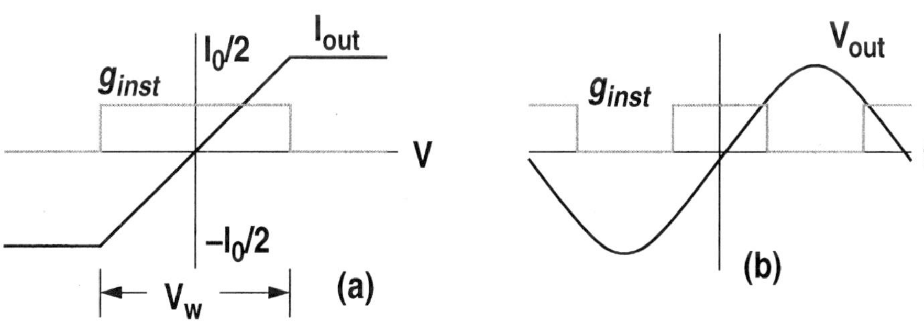

A reactive current, injected by the differential pair or the limiting amplifier, is captured by the quadrature terms given by g_{2nQ}. For the ideal differential pair, there is no reactive current so the quadrature terms are equal to zero.

The amplitude of the fundamental of the output current is derived using (6). This term is written as $I_c(t)$. Since:

$$\frac{dI}{dt} = \frac{dI}{dV}\frac{dV}{dt} = g(V)\frac{dV}{dt}, \tag{9}$$

$I_c(t)$ may be written as

$$I_c(t) = I(V_c(t)) = \int g(V_c(t))\frac{dV_c}{dt}dt. \tag{10}$$

This equation shows that mixing between the transconduction function and the derivative causes different terms of $g(V)$ to fall at ω_o. For example, the second harmonic of $g(V)$ mixes with the carrier to produce components at ω_o and $3\omega_o$. For oscillators, only the terms at ω_o are of interest since the resonator attenuates any out of band tones.

High Q resonators drastically attenuate higher harmonics. For this case, a final value of $I_c(t)$ is easily calculated and is given below. The transconductance terms are combined and referred to as the *effective* transconductance.

$$I_c(t) = \left(g_0 - \frac{g_{2I}}{2}\right)V_1\cos(\omega_o t) = g_{eff}V_1\cos(\omega_o t) \tag{11}$$

The intermodulation tones that result from the lower sideband, V_{lsb}, are described by the second term of (6). The resulting current tones that fall near the carrier are given as

$$I_{m1} = g_0 V_1 \cos((\omega_o - \omega_m)t + \phi) + \frac{g_{2I}}{2}V_1\cos((\omega_o + \omega_m)t - \phi). \tag{12}$$

If the lower sideband was part of a phase modulation pair, then the upper sideband, V_{usb}, must be

$$V_{usb}(t) = -V_m\cos((\omega_o - \omega_m)t - \phi). \tag{13}$$

The resulting current tones from this side band are

$$I_{m2} = \frac{-g_{2I}}{2}V_1\cos((\omega_o - \omega_m)t + \phi) - g_0 V_1\cos((\omega_o + \omega_m)t - \phi). \tag{14}$$

The total output sidebands are given as

$$I_{mt} = V_1\left(g_0 - \frac{g_{2I}}{2}\right)(\cos((\omega_o - \omega_m)t + \phi) - \cos((\omega_o + \omega_m)t - \phi)). \tag{15}$$

The output carrier-to-sideband ratio is easily calculated by using (11) and (15),

$$\frac{I_1}{I_{mt}} = \frac{\left(g_0 - \frac{g_{21}}{2}\right)V_1}{\left(g_0 - \frac{g_{21}}{2}\right)V_m} = \frac{V_1}{V_m}. \tag{16}$$

This completes the proof that in a nonlinear element, the ratio of the carrier to the PM sidebands at the input is equal to the output.

Based on this property, if a pair of phase modulation sidebands is applied to a limiting amplifier, the amount of phase modulation remains unchanged. Therefore, in a transimpedance limiting amplifier, the carrier-to-sideband ratio of a voltage input is equal to the carrier-to-noise ratio of the current output. An example of this circuit is the differential pair in the switched current oscillator, Figure 7. If it is driven by an oscillation on which phase noise sidebands are superimposed, the output current carrier-to-noise ratio is the same as the carrier-to-noise ratio of the input. This fact greatly simplifies the phase noise analysis. Using this simple property, the various sources of noise can be easily analyzed beginning with noise in the resonator.

FIGURE 7 *SNR of phase modulated signal is preserved at the output of an ideal limiter. (b) The differential pair is an example of a limiter.*

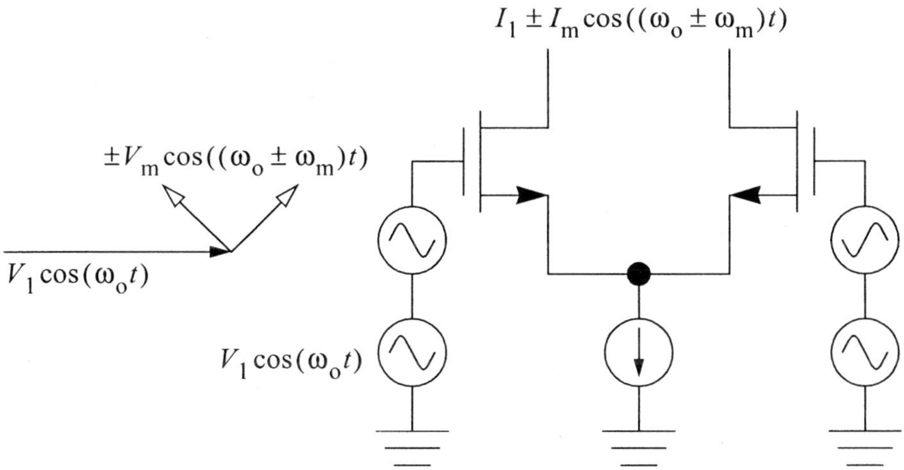

3.1 Resonator Noise

Figure 8a shows a simplified model of the cross-coupled oscillator. The resistance models the loss in the resonator. Also shown is the associated thermal noise current. Any noise can be decomposed into a series of sinewaves all uncorrelated in phase, Figure 8b [6]. The analysis begins with a sinewave component of this noise just above the oscillation frequency, Figure 8c. This can be decomposed into equal parts of PM and AM sidebands.

FIGURE 8 *Noise in resonator is modeled as a series of sinewaves. One component, just above the oscillation frequency, is decomposed into equal parts of PM and AM noise.*

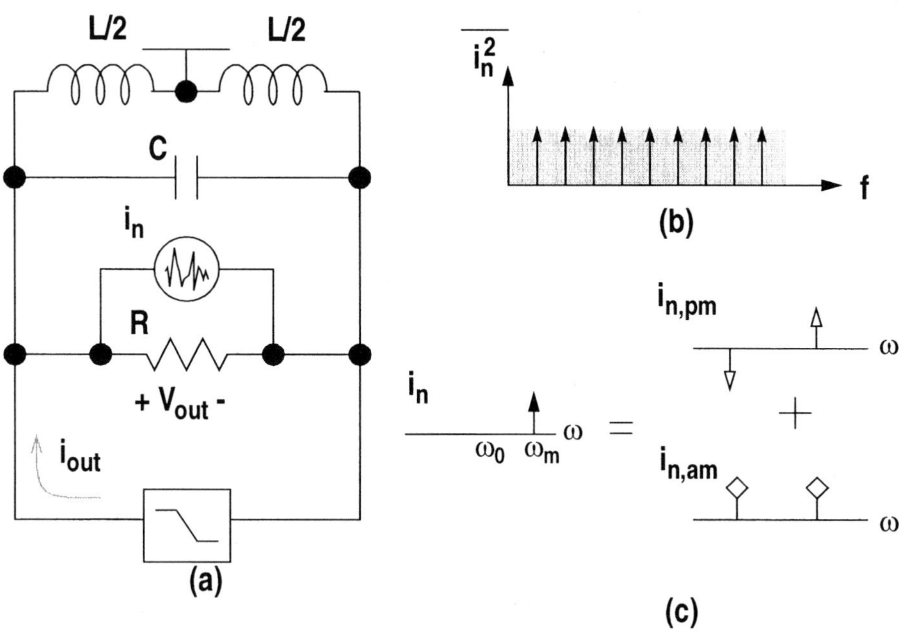

First, assume that in steady state there is an oscillation voltage across the resonator. Surrounding it are some unknown PM and AM sidebands shown by the frequency spectrum in Figure 9a. In general, these are unequal and each would have an arbitrary phase. The PM sidebands have a level of a and the AM sidebands have a level of b. The output voltage is

FIGURE 9 *Assumed steady-state voltage and the response of the active circuit.*

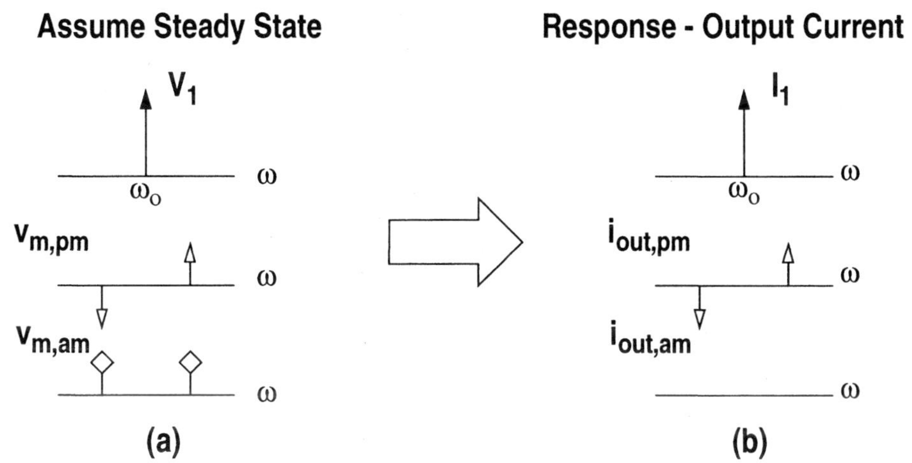

$$V_{out} = V_1 e^{j\omega t} + (a e^{j\omega_+ t} - a^* e^{j\omega_- t}) + (b e^{j\omega_+ t} - b^* e^{j\omega_- t}). \tag{17}$$

The cross coupled differential pair will react to this voltage and produce a current. However, as was shown earlier, this output will not contain AM, Figure 9b. The output is calculated using (16).

$$I_{out} = \frac{2}{\pi} I_0 e^{j\omega t} + \frac{2}{\pi} I_0 \left(\frac{a}{V_1} e^{j\omega_+ t} - \frac{a^*}{V_1} e^{j\omega_- t} \right) \tag{18}$$

In parallel with the resonator loss resistor is a noise current source. Only a single 1 Hz component of this current source located above the oscillation frequency is analyzed. This component is described by

$$i_{n,r} = i_n e^{j\omega_+ t}. \tag{19}$$

Next, KCL is performed on the left side of the resonator. Each frequency component is summed separately beginning with the tones at the oscillation frequency:

$$\frac{2}{\pi} I_0 = \frac{V_1}{R}. \tag{20}$$

At ω_+:

$$\frac{2}{\pi}I_0\frac{a}{V_1} = (a+b)\left[\frac{1}{R}+\frac{1}{jL\omega_+}+jC\omega_+\right]+i_n. \tag{21}$$

At ω_-:

$$-\frac{2}{\pi}I_0\frac{a^*}{V_1} = (-a^*+b^*)\left[\frac{1}{R}+\frac{1}{jL\omega_-}+jC\omega_-\right]. \tag{22}$$

Take complex conjugate of (22):

$$-\frac{2}{\pi}I_0\frac{a}{V_1} = (-a+b)\left[\frac{1}{R}-\left(\frac{1}{jL\omega_-}-jC\omega_-\right)\right]. \tag{23}$$

Next, subtract (23) from (21),

$$\frac{4}{\pi}I_0\frac{a}{V_1} = a\left[\frac{2}{R}+\frac{1}{jL\omega_+}-\frac{1}{jL\omega_-}+jC\omega_+-jC\omega_-\right]+ \tag{24}$$

$$b\left[\frac{1}{jL\omega_+}+\frac{1}{jL\omega_-}+jC\omega_++jC\omega_-\right]+i_n.$$

It is easy to show the following approximations:

$$\frac{1}{jL\omega_+}-\frac{1}{jL\omega_-}+jC\omega_+-jC\omega_- > 2\frac{\omega_m}{\omega_o}\left(j\omega C-\frac{1}{j\omega L}\right) \tag{25}$$

and

$$\frac{1}{jL\omega_+}+\frac{1}{jL\omega_-}+jC\omega_++jC\omega_- > 2\left(j\omega C+\frac{1}{j\omega L}\right). \tag{26}$$

Substituting these approximations into (24) results in

$$\frac{4}{\pi}I_0\frac{a}{V_1} = 2a\left[\frac{1}{R}+\frac{\omega_m}{\omega_o}\left(j\omega C-\frac{1}{j\omega L}\right)\right]+2b\left(j\omega C+\frac{1}{j\omega L}\right)+i_n. \tag{27}$$

By using (20) and assuming that the PM sidebands are much larger than the AM sidebands leads to the gain from a noise current to PM sidebands.

$$a = \frac{i_n}{2}\frac{\omega_o}{\omega_m}\frac{j}{\left(j\omega_o C + \frac{1}{j\omega_o L}\right)} = i_n \frac{j\omega_o L}{4}\frac{\omega_o}{\omega_m} \tag{28}$$

The phase noise of the oscillator due to the loss in the resonator is easily calculated by including noise from the lower sideband and normalizing with the RMS value of the amplitude of output oscillation. Note that the power of the resistor noise current is *4kT/R*.

$$\mathcal{L}(\omega_m) = \frac{2a^2}{V_1^2/2} = \frac{4}{V_1^2}\frac{4kT}{R}\left(\frac{\omega_o L}{4}\right)^2\left(\frac{\omega_o}{\omega_m}\right)^2 \tag{29}$$

With the final expression simplified as

$$\mathcal{L}(\omega_m) = \frac{4kTR}{V_1^2}\left(\frac{\omega_o}{2Q\omega_m}\right)^2. \tag{30}$$

To understand visually what is happening, a simple model is derived by separating the resonator noise current into PM and AM components:

$$i_{n,r} = \left[\frac{i_n}{2}e^{j\omega_+ t} - \frac{i_n^*}{2}e^{j\omega_- t}\right] + \left[\frac{i_n}{2}e^{j\omega_+ t} + \frac{i_n^*}{2}e^{j\omega_- t}\right]. \tag{31}$$

If only PM noise is injected into the resonator, the analysis is identical to that just performed. The results show that this is equivalent to injecting a PM noise current into a lossless tank, as shown in Figure 10. The PM component of the noise sinewave will now flow in this lossless resonator and be shaped, decaying inversely with frequency. The gain from the noise tone to the phase noise at the output is simply half the noise amplitude (to account for the PM component) times the impedance of a lossless tank. This is consistent with modeling the cross coupled differential pair as a negative resistance that exactly cancels the loss resistor at the carrier and PM sidebands at all other frequencies.

$$v_{o,n} = i_n \frac{L\omega_o^2}{4\omega_m} \tag{32}$$

If only AM noise is injected, the analysis is slightly different. Following the same steps as before, KCL at the upper and lower sidebands is given below.

FIGURE 10 *For the PM component of the noise, (a) the active circuit cancels the resistor and all the noise current flows into a (b) lossless resonator.*

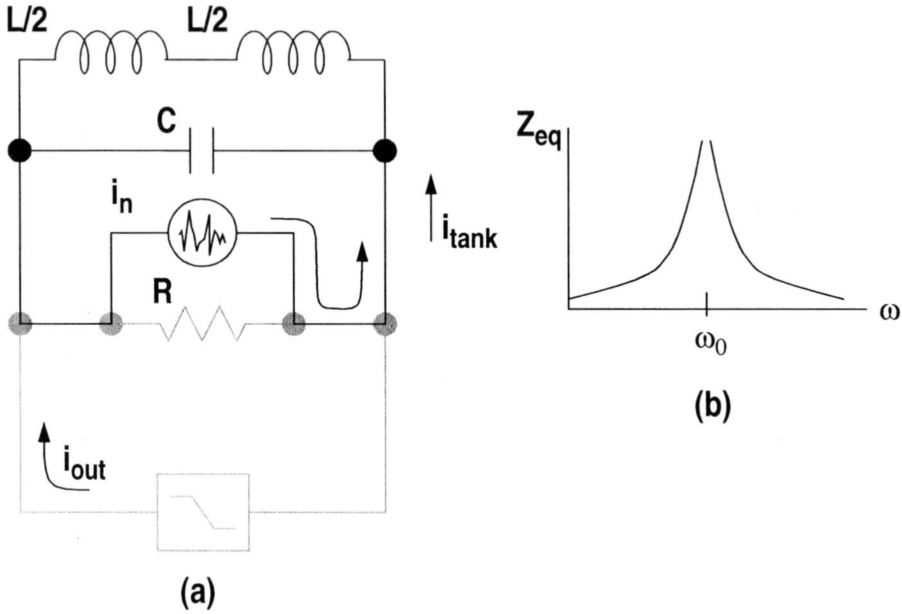

$$\omega_-: -\frac{2}{\pi} I_o \frac{a^*}{V_1} = (-a^* + b^*)\left[\frac{1}{R} + \frac{1}{j\omega_- L} + j\omega_- C\right] + \frac{i_n^*}{2}. \tag{33}$$

$$\omega_+: \frac{2}{\pi} I_o \frac{a}{V_1} = (a + b)\left[\frac{1}{R} + \frac{1}{j\omega_+ L} + j\omega_+ C\right] + \frac{i_n}{2}. \tag{34}$$

When the complex conjugate is taken of the lower sideband and the two equations are subtracted from each other, the noise current is canceled, leaving

$$0 = a\left[-\frac{1}{j\omega_- L} + \frac{1}{j\omega_+ L} - j\omega_- C + j\omega_+ C\right] + b\left[\frac{1}{j\omega_- L} + \frac{1}{j\omega_+ L} + j\omega_- C + j\omega_+ C\right]. \tag{35}$$

The approximations given before reduce this equation to

$$0 = a\left(\frac{\omega_m}{\omega_o}\right)\left[\frac{-1}{j\omega_o L} + j\omega_o C\right] + b\left[\frac{1}{j\omega_o L} + j\omega_o C\right]. \tag{36}$$

If oscillation occurs near the resonant frequency of the oscillator, the second term in (36) is equal to zero. This results in $a = 0$, or no PM due to the AM current. Setting $a = 0$ from above into (33) and (34):

$$\omega\_: -\frac{i_n^*}{2} = b^*\left[\frac{1}{R} + \frac{1}{j\omega\_L} + j\omega\_C\right], \tag{37}$$

$$\omega_+: \frac{i_n}{2} = b\left[\frac{1}{R} + \frac{1}{j\omega_+ L} + j\omega_+ C\right]. \tag{38}$$

Equations (37) and (38) show the result is the same if the AM noise was driven into the lossy tank. This is shown in Figure 11. (39)

FIGURE 11 *For the AM component of the noise, the active circuit cancels the resistor and the noise flows into the lossless resonator and the resistor.*

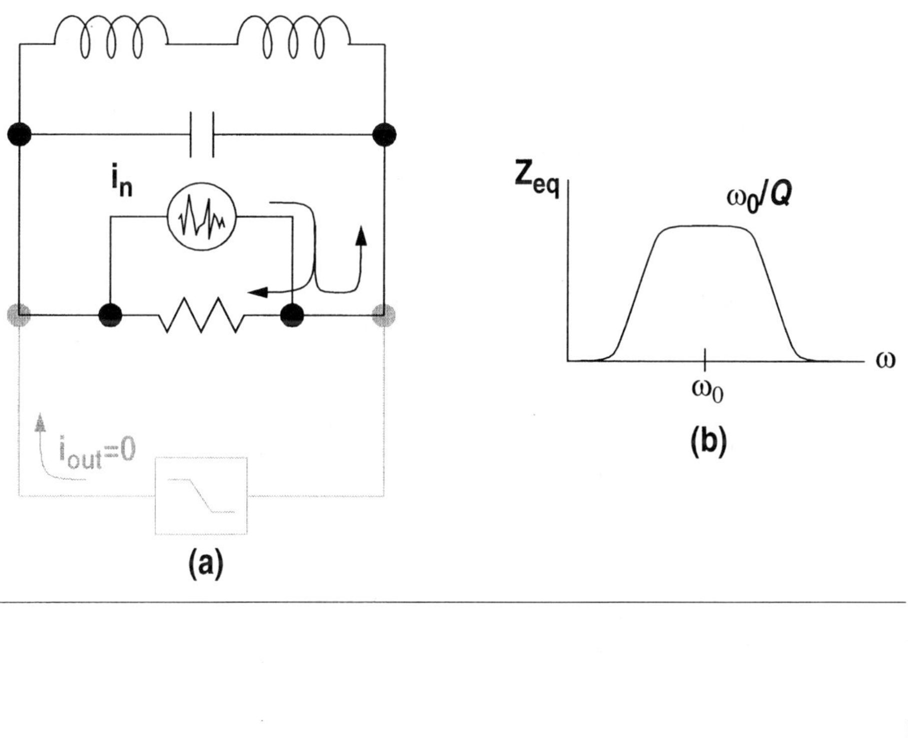

This circuit was simulated in SpectreRF as a function of tail current and with various combinations of transistor size and quality factors. As the current was increased, the resistance was decreased to keep the amplitude of the carrier constant. The complete simulation parameters are given in Table 1. From that, simulated gain from the noise current to phase noise at the output is plotted in Figure 13b. The expression derived in (32) matches the simulations exactly.

TABLE 1 *Simulation parameters for the results given in Figure 12: L = 100nH, C=1pF,* f_m *= 10 Hz.*

Curve	*IR*	$k = \frac{\mu C_{ox}}{2}\frac{W}{L}$
1	1 V	0.001
2	1 V	0.005
3	2 V	0.001
4	2 V	0.005

In summary, while noise near the oscillation frequency is important, the resonator filters noise farther away, Figure 13a. Furthermore, the phase noise expression shows that as the quality factor of the resonator increases, the oscillation amplitude increases as Q-squared before the supply limit, whereas the sidebands will go down as Q. As a result, the total phase noise improves as $1/Q$-cubed, Figure 13b.

3.2 Differential Pair Noise

Next, the noise in the differential pair transistors is analyzed. Noise is injected into the resonator only when the differential pair is in its active region, Figure 14a. For example, if one switch is *off*, it obviously contributes no noise, and neither does the other switch that is *on* because it acts as a cascode transistor whose tail current is fixed to I_0 by the current source. So, if the sinewave in Figure 14b drives the differential pair, pulses of noise carrying current appear at the output of the differential pair.

An expression for the output current for a single component of noise located just above the carrier is easily derived. Using the fact that the PM SNR is pre-

FIGURE 12 *Simulation results of "phase noise gain" versus bias current, resonator loss, and transistor transconductance. (32) has excellent agreement with the simulation results. See Table 1 for the circuit parameters.*

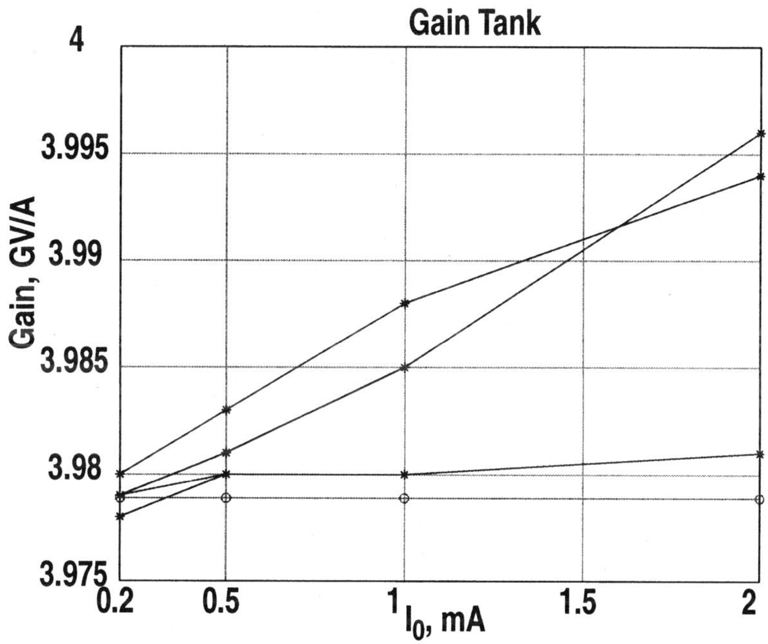

served in a nonlinear device and that a noise tone can be decomposed into half AM and half PM, the output noise is given in (40).

$$\frac{i_n}{I_1} = \frac{v_{n,pm}}{V_1} \rightarrow i_n = \frac{v_n}{2R} \tag{40}$$

Multiplying this expression by the impedance of a lossless tank results in the gain from a single noise component to phase noise.

$$v_{o,n} = v_n \frac{L\omega_o^2}{4R\omega_m} \tag{41}$$

Again, SpectreRF simulations give the gain from the noise voltage to the phase noise at the output and are plotted for swept tail current and for various

FIGURE 13 *(a) Noise far from the carrier is filtered by the tank. (b) As the Q of the resonator increases, the amplitude increases by Q-squared and the noise level drops by Q, resulting in a Q-cubed improvement.*

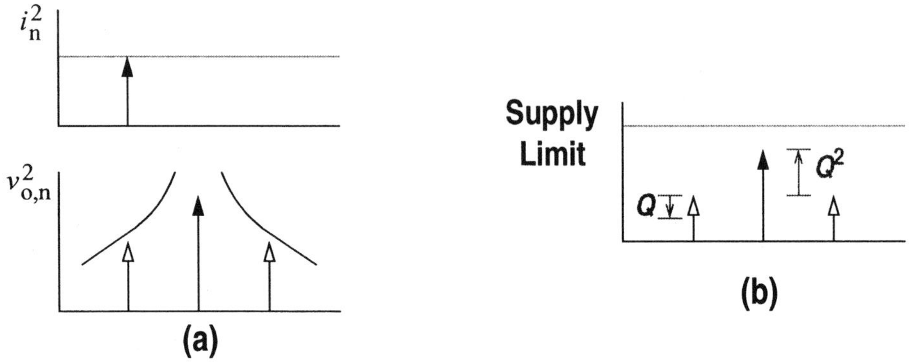

other conditions in the oscillator, with all the values appearing in Table 1. The expression in (41) exactly predicts the simulation results, Figure 15.

The complete noise mechanism is based on the mixer noise paper by Darabi [11]. His analysis is more rigorous but is equivalent to that presented here. As stated before, noise only appears at the output when the devices are in the active region. At this time, the transconductance of the differential current, the current that actually flows through the resonator, is $G_m = I_0/V_w$, Figure 6a. This is half the transconductance of the individual devices, (42). This important difference is due to only the differential current flowing through the resonator.

$$G_m = \frac{I_0}{V_w} = \frac{g_{m_{1,2}}}{2} \tag{42}$$

As oscillation occurs, the instantaneous transconductance toggles with time from G_m to 0, Figure 6b. The noise current that appears at the output is given in (43) where v_n is the input referred voltage noise and $G_m(t)$ is the transconductance envelope.

$$i_{o,n} = v_n G_m(t) \tag{43}$$

FIGURE 14 *Noise in the switching devices is injected into the resonator as pulses only when the transistors are in the linear region. These noise pulses have a width, t_w*

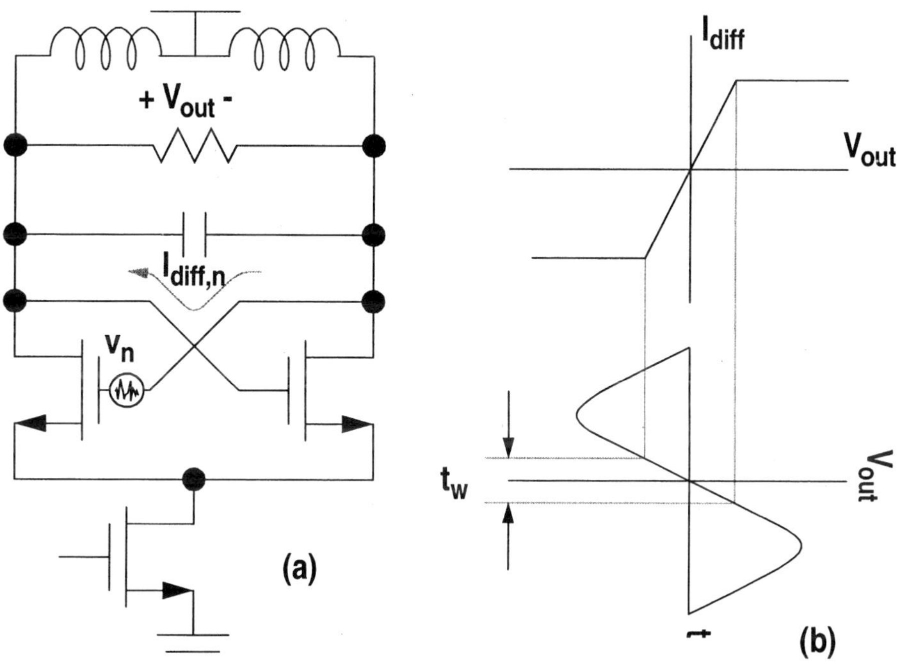

Since the transconductance is varying with time, the output noise appears as pulses of noise current with a pulse width equal to the duration the differential pair is in the linear region. Figure 16a shows the envelope of the noise pulses in the time domain and Figure 16b shows them in the frequency domain. In the frequency domain, the envelope is shaped with a $\sin(x)/x$ envelope with the first null at one over t_W where t_W is the pulse width. An important characteristic is the pulses only appear at even harmonics of the oscillation frequency.

This mixing process causes noise folding to occur, allowing noise from many frequency locations to congregate at one point near the oscillation frequency. The frequency spectrum of the noise pulses is convolved with the white noise density of the noise voltage, Figure 16c, to produce the output noise, Figure 16d [10]. To understand process, the convolution is examined one

FIGURE 15 *Simulated comparison of the phase noise gain from a noise component near the oscillation frequency. Simulations agree with (41).*

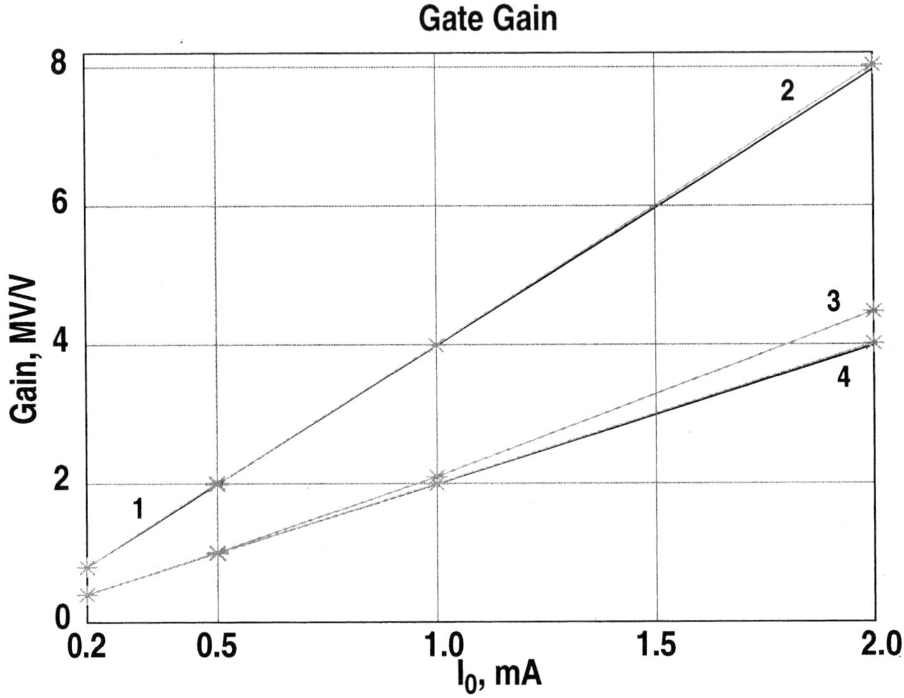

component at a time. Suppose there is noise just above the oscillation frequency (labeled "1" in Figure 16c), this convolves with the first two components of the noise pulses located at DC and $2\omega_0$ and produces a pair of sidebands around the carrier. These sidebands have a phase relationship that shows they are phase noise. Further, noise located below the oscillation frequency will likewise produce phase noise sidebands around the carrier. Next, consider noise near the second harmonic (labeled "2" in Figure 16c). This convolves with the first three components of the noise pulses (DC, $2\omega_o$, and $4\omega_o$) to produce phase noise sidebands, which are far away from the oscillation frequency. Finally, noise at the third harmonic convolves with the second and third component to again produce phase noise around the carrier. From this example it is clear that noise at the oscillation frequency and at odd multiples is important. Also, the noise is eventually band-limited by the sinc envelope.

FIGURE 16 *Noise in the differential pair appears as noise pulse in the time domain, (a). In the frequency domain, these pulses are the result of a convolution between a sin(x)/x function, (b), and the white noise PSD of the transistor, (c). Only noise near odd harmonics fall within the bandwidth of the resonator.*

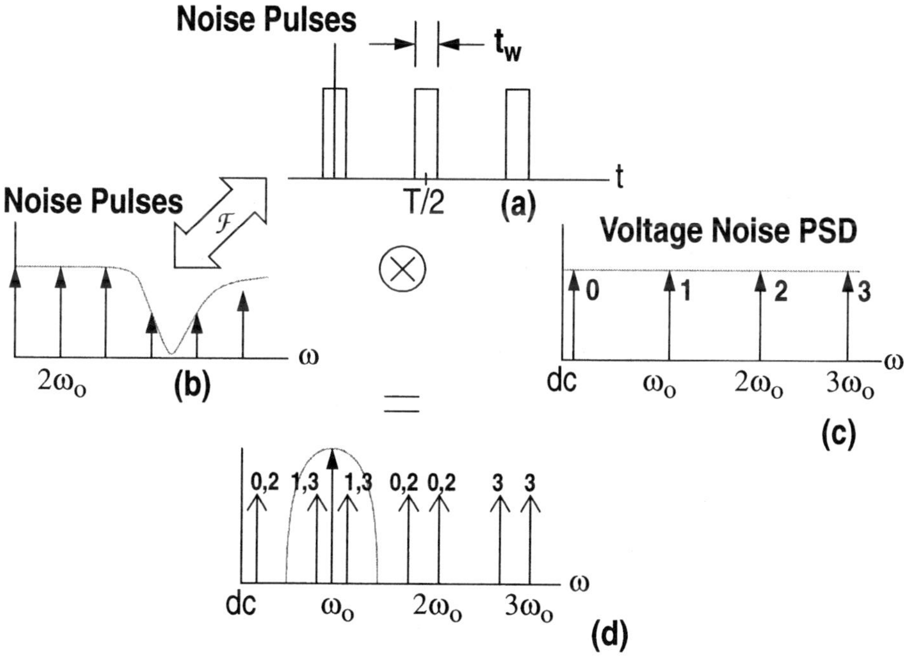

There are a number of ways to calculate the summation of the convolution terms as the $\sin(x)/x$ function decays. Darabi presents a theoretically rigorous method. A simplified method is presented here that produces the same result as Darabi but is easier to visualize. To simplify the analysis, the $\sin(x)/x$ function in the frequency domain is approximated as having impulses with constant amplitude up to $1/t_w$ and zero elsewhere, Figure 17.

The DC component, $g[0]$ is simply the DC value of the time-domain waveform.

FIGURE 17 *Noise pulses are approximated as having constant amplitude up to the null of the sin(x)/x function.*

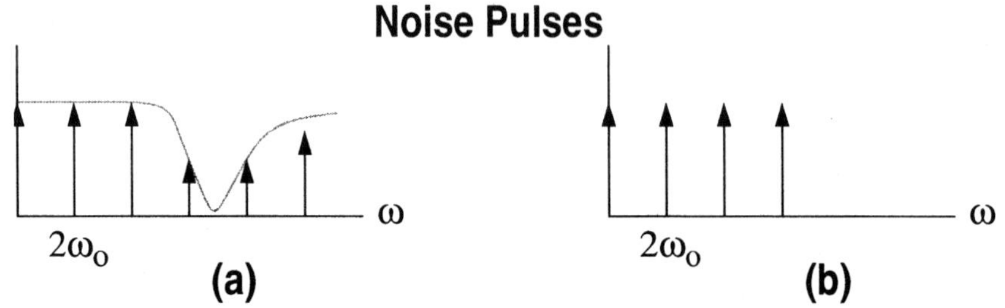

$$g[0] = G_m \frac{t_w}{T/2} = 2G_m \frac{t_w}{T} \tag{44}$$

Since all the non-zero terms are equal to $g[0]$, the frequency domain function is fully described.

To verify this value, the gain from a noise located above the carrier to output noise is again calculated. Recall that the DC term mixes with the upper sideband of the voltage noise to produce an output current located at the upper sideband. This is shown in (45) which matches (40).

$$i_{n,pm} = v_{n,usb} g[0] = v_{n,usb} 2 \frac{I_0}{V_w} \frac{V_w}{V_1 \omega_o} \frac{1}{T} = \frac{v_{n,usb}}{2R} \tag{45}$$

Since each frequency term convolves with white noise, the final answer is simply the sum of the square of the frequency domain envelope times the white noise. Since all the frequency terms are equal, only the number is required. This is just the bandwidth divided by $2f_o$: $N_{fold} = (1/t_w)/(2f_o) = 1/2f_o t_w$. Equation 46 gives the final output noise where N_{dev} is the two devices in the differential pair.

$$i_{o,n}^2 = \frac{v_n^2}{4R^2} N_{fold} N_{dev} \tag{46}$$

The equivalent voltage noise is set by the transconductance of the individual devices,

$$v_n^2 = \frac{4kT}{g_{m_{1,2}}}. \tag{47}$$

After substituting for the value of the voltage noise, the folding term, and the number of devices, the final output noise is calculated,

$$i_{o,n}^2 = 4kT\gamma\frac{I_0}{\pi V_1}. \tag{48}$$

This equation shows that the output noise density at the switches only depends on the amplitude and the bias current, and not on the transistor size!

The constant noise is because the noise in the differential pair is not sampled by impulses, but by time windows of finite width. The window height is proportional to transconductance, and the width is set by the tail current and the slope of the oscillation waveform at the zero crossing. The input-referred noise spectral density of the differential pair is inversely proportional to transconductance. Thus, the narrower the sampling window — that is, the larger the sampling bandwidth — the lower the noise spectral density [11].

The final phase noise due to the switching devices is calculated by scaling the output noise by the loss tank and normalizing it by the amplitude of the fundamental.

$$\mathcal{L}(f_m) = \frac{8kT\gamma I_0}{\pi V_1^3}R^2\left(\frac{\omega_o}{2Q\omega_m}\right)^2 \tag{49}$$

Again, the total phase noise in (49) depends only on the current, the resistance, and the amplitude of oscillation and is independent of the size or the transconductance of these devices. Also, as the quality factor improves, we get a $1/Q$-cubed improvement in phase noise below the supply-limited amplitude, Figure 18.

3.3 Tail Current Noise

Finally, the third source of noise is in the tail current, Figure 19. Any noise in this transistor is commutated and frequency translated by the switching pair as in a single balanced mixer and injected into the tank. It is assumed that the amplitude of oscillation is much larger than the transition region of the differ-

FIGURE 18 *Once again, as Q increases, the amplitude increases at Q-squared while the sidebands are attenuated by Q, resulting in a Q-cubed improvement in phase noise.*

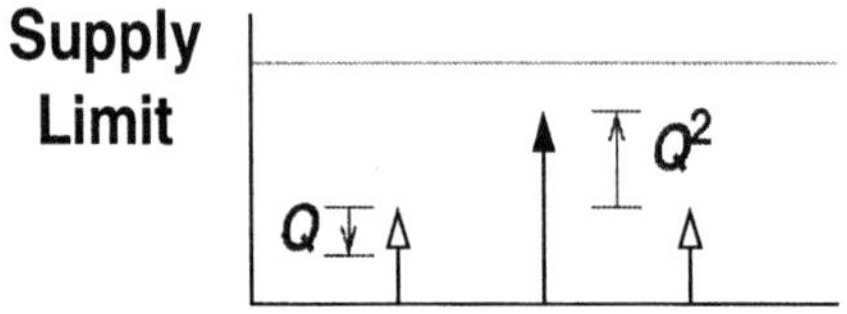

ential pair. When this is the case, the mixing function is more of a square wave than a sinewave.

FIGURE 19 *Noise in the current source is the third source of phase noise.*

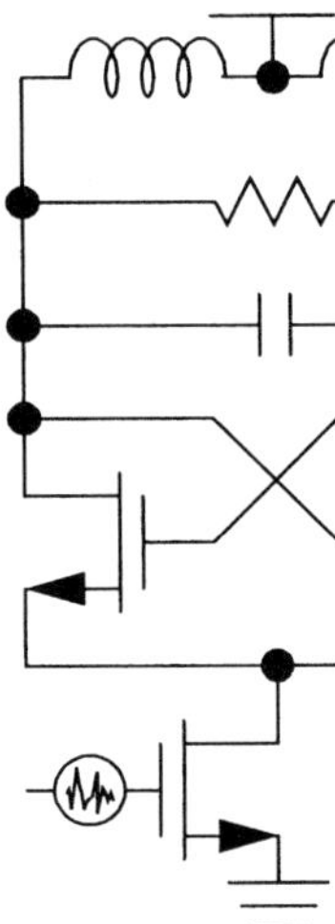

Therefore, noise at low frequencies is mixed up into a pair of sidebands around the carrier and injected into the tank. However, these are AM sidebands and therefore not important, Figure 20. Any varactor connected to the resonator will convert the AM envelope into FM and that will produce phase noise [12]. Noise located at the oscillation frequency produces sidebands at low frequencies and at the second harmonic, which are far away from the carrier.

Noise originating at the second harmonic is converted around the carrier and around the third harmonic. The noise translated around the third harmonic is attenuated by the resonator and is ignored. The noise that is translated near the carrier arrives there from two components of the mixing function. The fundamental component of the square wave will down convert the noise at the second harmonic to one side of the carrier while the third harmonic of the square wave will down convert the noise to the other side of the carrier. However, the third harmonic has a conversion gain of 1/3 of the fundamental so the mixing process is not symmetric.

The resulting noise is decomposed in a third that appears as AM while the remaining two-thirds appear as a single sideband noise that lands in-band. So, in the absence of varactors, the only noise that produces phase noise is that which lies at the second harmonic and all even multiples.

FIGURE 20 *Only noise located at even harmonics will translate in-band and produce phase noise.*

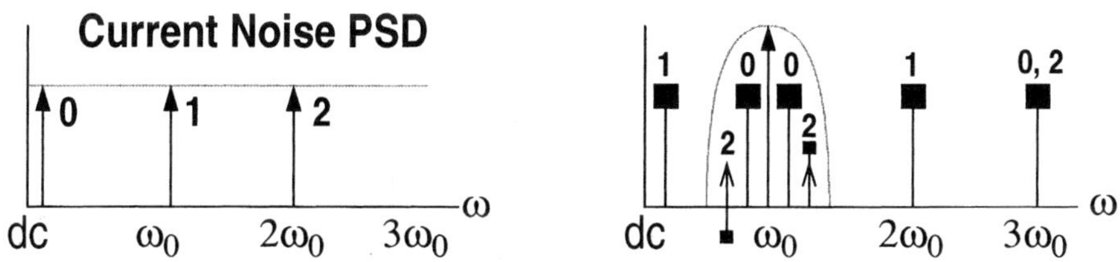

For example, the noise is modeled as $i_n = \sin(2\omega_0+\omega_m)t$ and the square wave mixing function is approximated in (50). If the square wave function is modeled with a cosine, the third harmonic is negative to allow the zero crossings of the harmonic to align.

$$m(t) = \frac{2}{\pi}\left[\sin\omega_o t + \frac{1}{3}\sin 3\omega_o t + \dots\right] \tag{50}$$

The terms of the product between the noise and the mixing function are truncated to the frequency of interest and are given in

$$i_{out}(t) = \frac{i_n}{\pi}\left[\cos(\omega_o + \omega_m)t + \frac{1}{3}\cos(\omega_o - \omega_m)t + \dots\right]. \tag{51}$$

These terms are separated into PM and AM components.

$$i_{out}(t) = \frac{i_n}{\pi}\left(\frac{1}{2}+\frac{1}{6}\right)[\cos(\omega_o+\omega_m)t+\cos(\omega_o-\omega_m)t] + \frac{i_n}{\pi}\left(\frac{1}{2}-\frac{1}{6}\right)[\cos(\omega_o+\omega_m)t-\cos(\omega_o-\omega_m)t] \quad (52)$$

Equation 53 below describes the resulting phase noise sidebands corresponding to the second harmonic input. It is simply the PM sideband in (52) times the impedance of the lossless tank. Again, SpectreRF simulations are shown for this transfer function as a function of tail current and various other parameters in the oscillator. And again, a good fit is obtained for this equation, Figure 21.

$$v_{o,n} = \frac{2i_n}{3\pi}\frac{L\omega_o^2}{2\omega_m} \quad (53)$$

The small departures from (53) arise from the fact that the frequency conversion process is sensitive to the square-wave mixing that was used to derive this equation. Since the commutation will often be less than a square-wave, (53) is pessimistic as Figure 21 clearly shows.

To calculate the total noise, all the noise locations must be included. The noise located on the other side of the second harmonic contributes an equal amount of noise. Since the square-wave mixing process is rich in harmonics, noise from higher harmonics will all be down converted to near the oscillation frequency. Since all these noise sources are uncorrelated, they must be summed powerwise. Additionally, only the PM component is included.

$$\hat{v}_p^2 = \frac{1}{2\pi^2}\left[\left(\frac{1}{1}+\frac{1}{3}\right)^2+\left(\frac{1}{3}+\frac{1}{5}\right)^2+\ldots\right]\left(\frac{L\omega_o}{2}\right)^2\left(\frac{\omega_o}{\omega_m}\right)^2\hat{i}_n^2 \quad (54)$$

The summation conveniently sums to $\pi^2/4$. The simplified expression is given in

$$\hat{v}_p^2 = \frac{1}{8}\left(\frac{L\omega_o}{2}\right)^2\left(\frac{\omega_o}{\omega_m}\right)^2\hat{i}_n^2. \quad (55)$$

FIGURE 21 *Simulation vs. equation for the transfer function from 2nd harmonic in the current source to phase noise sidebands.*

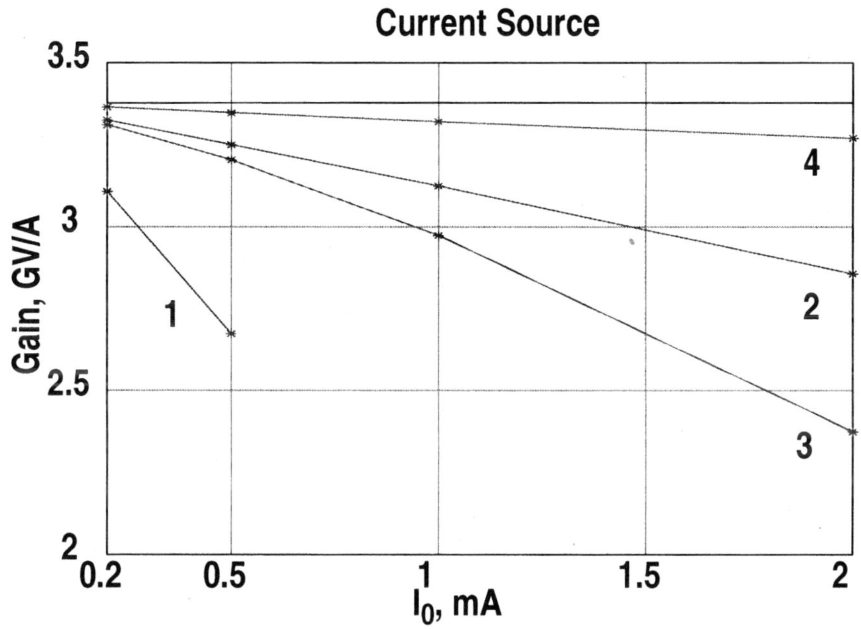

Normalizing (55) results in the expression for total phase noise, including all the important harmonics.

$$\mathcal{L}(f_m) = \frac{kT\gamma\pi^2}{2I_0V_{eff}}\left(\frac{\omega_o}{2Q\omega_m}\right)^2 \tag{56}$$

Where γ is the noise factor of a single FET, classically 2/3. Here, as the quality factor improves, the amplitude rises as Q squared but the sideband levels remain constant. As a result, the improvement in phase noise is $1/Q$-squared, Figure 22.

It is important to note that the AM noise resulting from up conversion, if impressed across a varactor at the resonator, will modulate the varactor, resulting in AM-to-FM conversion [12,13]. Although the process is different, the sidebands are indistinguishable from PM noise sidebands. Unlike the

FIGURE 22 *As Q changes, only the amplitude increases, resulting in a Q-squared improvement in phase noise.*

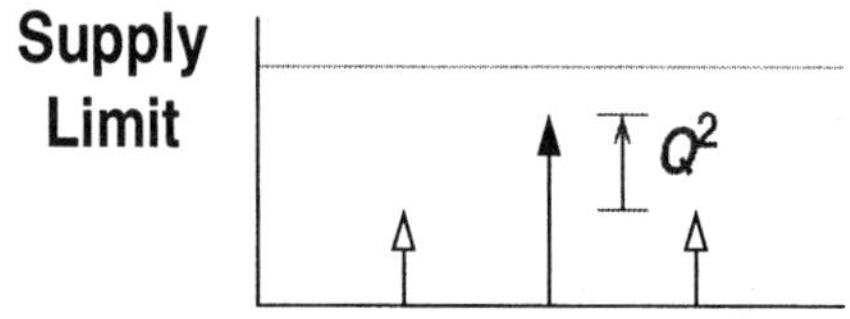

other mechanisms of phase noise, this effect depends on the varactor characteristics and VCO tuning range and it may be significant only in certain situations.

The total phase noise is derived by summing

$$\mathcal{L}(\omega_m) = \frac{4kTR}{V_1^2}\left[1 + \frac{2\gamma I_0 R}{\pi V_1} + \frac{\gamma I_0 R}{2V_{eff}}\right]\left(\frac{\omega_o}{2Q\omega_m}\right)^2 \tag{57}$$

3.4 Proving Leeson's Hypothesis

Recall Leeson's original equation from Section 2 contains a circuit specific factor *F*. Equation (57) derives a precise expression for this unknown noise factor and is given in (58). It contains the circuit specific parameters and various other terms, which are all well understood.

$$F = 1 + \frac{2\gamma IR}{\pi V_1} + \frac{\gamma IR}{2V_{eff}} = 1 + \gamma\left(1 + \frac{\pi V_1}{4V_{eff}}\right) \tag{58}$$

The differential oscillator it consists of three terms. The first is due to noise in the resonator. The second term is due to noise in the differential pair, which by the way, is independent of the differential pair transistors themselves. And the final term results from the current source noise. This completely specifies the phase noise in the white noise region.

Herzel ignored the noise due to the current source, and through simulation, derived a similar expression [14]. The results derived here are similar to those by Samori [3]. Samori's derivation is similar in spirit to that presented here but is much more mathematical and does not lead to the same intuitive understanding of the oscillator. Additionally, he was unable to obtain a closed-form expression for this last term.

The simple expression (58) captures all nonlinear effects and frequency translations. At low bias currents, while the amplitude of oscillation is smaller than the power supply, the differential pair acts as a pure current switch driving the resonator and $V_1 = (2/\pi)RI_0$ [15]. For this case, the second term comprising F simplifies to 2γ. This means that as tail current increases and assuming $g_{m,bias}R$ is held constant, the noise factor remains constant and phase noise improves as a function of I_0-squared. Others [15] have observed this. However, beyond a critical tail current the amplitude V_1 is pegged constant, limited by supply voltage. Further increases in I_0 will cause the differential pair's contribution to noise factor to rise, degrading phase noise proportional to I_0. Recall, the amplitude of oscillation rises with the tail current until it is limited by the supply voltage, Figure 3. From this expression, it is clear that phase noise decreases rapidly as the amplitude rises. Once it goes beyond the supply limit, the amplitude is pegged and the noise factor worsens with more bias current, degrading phase noise beyond that point. So there is a clear optimum point of operation for least phase noise. Therefore, to minimize phase noise, the tail current should be just enough to drive the amplitude to its maximum possible value.

4 Validation of Thermal Noise Analysis

Now, the total expression for phase noise was validated on two widely different differential oscillators. Figure 23 is the measured phase noise of two samples of an oscillator fabricated in 0.35μm CMOS with on-chip inductors with a low Q of 6, and oscillation frequency of 1.1 GHz and a tail current of 3.4 mA. It also contains a SpectreRF simulation of the phase noise.

Figure 24 is the measured phase noise of an oscillator fabricated in 0.25 μm CMOS with off-chip inductors with a high-Q of 25 but biased with a very low current of 0.3mA and an oscillation at 830 MHz.

Applying the formulas developed previously to these two oscillators and without using any fudge factors, excellent agreement is found in both cases within 1 dB of measurement. This serves as validation of these formulas.

FIGURE 23 *Measured and calculated phase noise of an oscillator with integrated inductors. Measurements from two chips are shown along with a SpectreRF simulation. The SpectreRF simulation error at low frequencies is due to a poor flicker noise model. The oscillation frequency is 1.1 GHz, the inductors have a Q of 6, and the circuit draws 3.4 mA.*

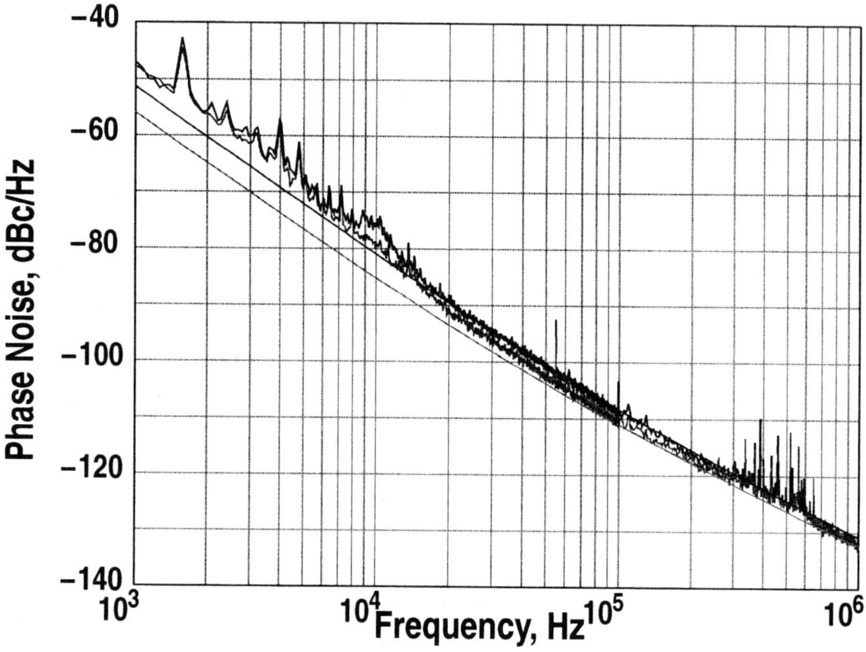

FIGURE 24 *Measured and calculated phase noise match well for a 0.25µm oscillator with off-chip inductors. The oscillation frequency is 830 MHz, the Q of the inductors is 25 and the circuit draws 0.3mA of current.*

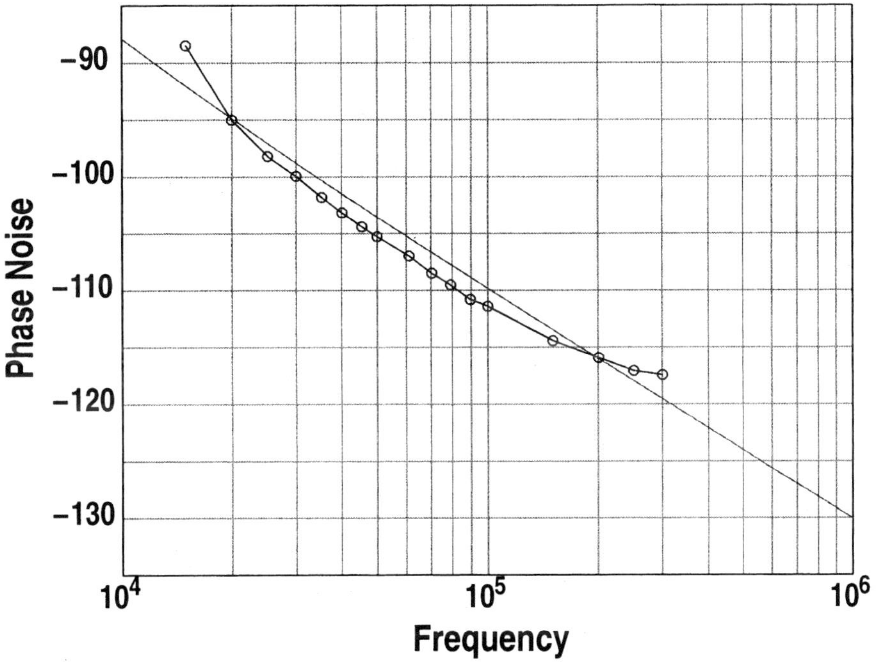

References

[1] A. Hajimiri and T. H. Lee, "A general theory of phase noise in electrical oscillators," *IEEE Journal of Solid-State Circuits*, vol. 33, pp. 179-94, 1998.

[2] J. Craninckx and M. Steyaert, "Low-noise voltage-controlled oscillators using enhanced LC-tanks," *IEEE Transactions on Circuits and Systems II: Analog and Digital Signal Processing*, vol. 42, pp. 794-804, 1995.

[3] C. Samori, A. L. Lacaita, F. Villa, and F. Zappa, "Spectrum folding and phase noise in LC tuned oscillators," *IEEE Transactions on Circuits and Systems II: Analog and Digital Signal Processing*, vol. 45, pp. 781-90, 1998.

[4] E. Hegazi, H. Sjoland, and A. A. Abidi, "A filtering technique to lower LC oscillator phase noise," in *Solid-State Circuits, IEEE Journal of*, vol. 36: Practical, 2001, pp. 1921-1930.

[5] D. B. Leeson, "A simple model of feedback oscillator noise spectrum," *Proceedings of the IEEE*, vol. 54, pp. 329-330, 1966.

[6] S. O. Rice, "Mathematical analysis of random noise," *Bell Systems Technical Journal*, vol. 23 and 24, 1945.

[7] G. Y. Belinkiy and B. Z. Kisel'gof, "The causes of parasitic PM in an amplitude limiter," *Telecommunications*, vol. 23, pp. 56-57, 1969.

[8] W. B. Davenport, "Signal-to-noise ratios in band-pass limiters," *Journal of Applied Physics*, vol. 24, pp. 720-727, 1953.

[9] V. A. Malyshev and G. G. Chervyakov, "Analysis of AM/PM conversion in arbitrary nonlinear elements," *Izvestiya Vysshikh Uchebnykh Zavedenii, Radioelektronika*, vol. 20, pp. 51-6, 1977.

[10] Joel Phillips and Ken Kundert, "Noise in mixers, oscillators, samplers, and logic: an introduction to cyclostationary noise," *www.designers-guide.com/Theory.*

[11] H. Darabi and A. A. Abidi, "Noise in RF-CMOS mixers: a simple physical model," *IEEE Journal of Solid-State Circuits*, vol. 35, pp. 15-25, 2000.

[12] E. Hegazi and A. A. Abidi, "Varactor characteristics, oscillator tuning curves, and AM-FM conversion," *IEEE Journal of Solid-State Circuits*, 2003.

[13] S. Levantino, C. Samori, A. Zanchi, and A. L. Lacaita, "AM-to-PM conversion in varactor-tuned oscillators," *Circuits and Systems II: Analog and Digital Signal Processing, IEEE Transactions on*, vol. 49, pp. 509-513, 2002.

[14] F. Herzel, M. Pierschel, P. Weger, and M. Tiebout, "Phase noise in a differential CMOS voltage-controlled oscillator for RF applications," *IEEE Transactions on Circuits and Systems II: Analog and Digital Signal Processing*, vol. 47, pp. 11-15, 2000.

[15] A. Hajimiri and T. H. Lee, "Phase noise in CMOS differential LC oscillators," presented at Symposium on VLSI Circuits, Hi, Usa, 1998.

4 Colpitts Oscillator

1 Introduction

The universe of oscillators can be divided between the differential LC oscillator, where the zero crossing of the oscillating voltage switches the active devices, and the Colpitts oscillator, where only the peaks of the oscillation voltage inject current from the active device.

Historically, when transistors were expensive, the Colpitts oscillator was the topology of choice because it required only a single transistor [1]. Now that transistors are essentially free, the Colpitts oscillator remains popular because it only requires a single pin to connect to an external resonator and no coupled inductors. However, its steady state behavior is poorly understood, and its phase noise is outright misunderstood.

Using complicated mathematical expressions, Huang analyzed the phase noise of the Colpitts oscillator in 1998 [2]. He gave a detailed analysis of the Colpitts oscillator and his methods served as a motivation for this work. In this chapter, the techniques developed in Chapter 3 are used to simplify the analysis of phase noise in the white noise region. As a historical note, Kulagin analyzed the Colpitts oscillator in a similar manner [1]. He specifically looked at a Colpitts oscillator with automatic gain control.

2 Steady-State

Figure 1 shows the Colpitts oscillator that is analyzed. This topology is exactly the same as the circuit analyzed by Huang except the location of the ground node has been moved to the source terminal. This change greatly simplifies the analysis.

The analysis begins with a large-signal steady-state analysis. Once oscillation starts, the capacitors charge up to a DC voltage which biases the transistor in a nominally off position. The amplitude builds up to a point such that the peaks of the oscillation voltage across capacitor C_1 turn on the transistor for a conduction angle, θ, which is a fraction of the phase of one period, 2π, Figure 2c. "Steady-state" implies the selection of θ such that various circuit constraints are met. First, the average current through the transistor must equal the bias current I_0 because the capacitors cannot carry DC current. Second, the fundamental component of the periodic transistor current, I_D, must support the amplitude across C_1 responsible for setting θ.

FIGURE 1 *Schematic of a Colpitts oscillator. The noise in the oscillator is modeled as a current source in parallel with the bias current source.*

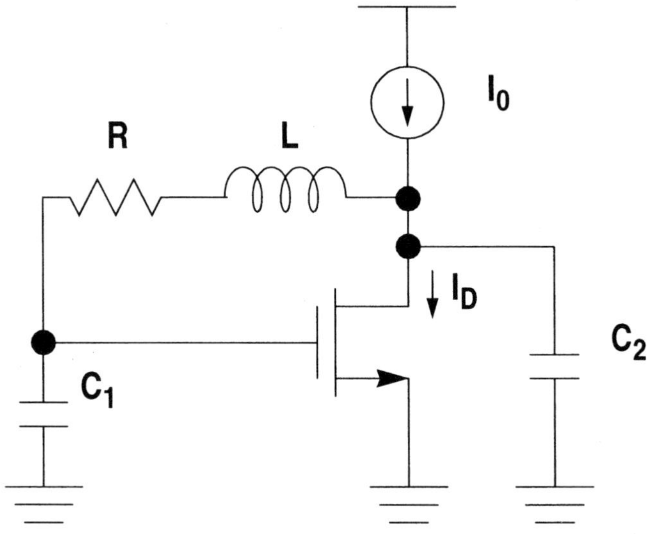

As in the analyses of other large-signal circuits, the transistor characteristics are simplified to capture its essential action. Here, the simplification entails representing it as a constant transconductance, g_m, above a threshold V_T, Figure 2a. The transistor carries an average current I_{D0} while its physics and/or its aspect ratio will determine g_m. Eventually g_m will be represented in terms of θ. To further simplify the analysis, the output voltage is taken as the voltage across C_1 and it is assumed to consist of an oscillation and DC component, (1).

$$V_{C1} = V_1 \cos(\omega_0 t) + V_0 \tag{1}$$

FIGURE 2 *(a) Ideal model of transistor in the Colpitts oscillator. (b) Transistor turns on only at the peak of oscillation. (c) The transistor output consists of tips of sinusoids.*

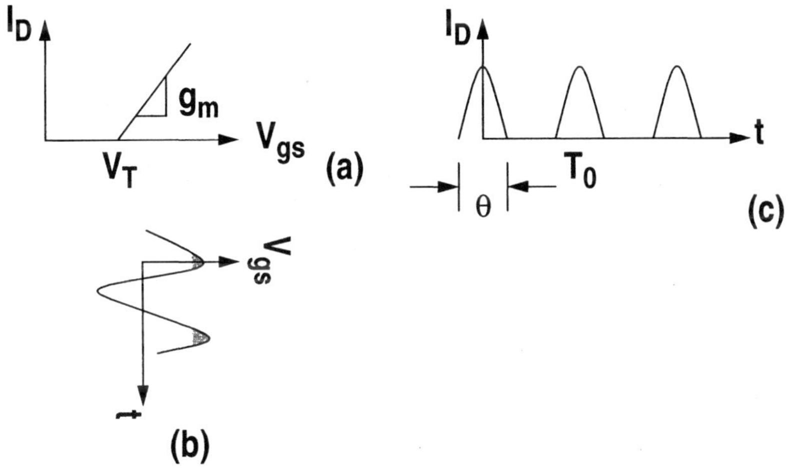

The transistor current waveform is given by:

$$I_D = \begin{cases} g_m V_1\left(\cos\phi - \cos\frac{\theta}{2}\right) & \text{for } -\frac{\theta}{2} < \phi < \frac{\theta}{2} \\ 0 & \text{for } -\pi < \phi < -\frac{\theta}{2}, \frac{\theta}{2} < \phi < \pi \end{cases} \tag{2}$$

where V_1 is the amplitude of sinewave on V_{C1} and $\phi = \omega_0 t$. The DC component of I_D is denoted I_{D0} and is

$$I_{D0} = \frac{g_m V_1}{2\pi} \int_{-\frac{\theta}{2}}^{\frac{\theta}{2}} \left(\cos\phi - \cos\frac{\theta}{2}\right) d\phi \tag{3}$$

$$I_{D0} = \frac{g_m V_1}{2\pi}\left(2\sin\frac{\theta}{2} - \theta\cos\frac{\theta}{2}\right) \tag{4}$$

The fundamental frequency component of I_D is denoted I_{D1} and is

$$I_{D1} = \frac{g_m V_1}{2\pi}\int_{-\frac{\theta}{2}}^{\frac{\theta}{2}}\left(\cos\phi - \cos\frac{\theta}{2}\right)\cos\phi\, d\phi \tag{5}$$

$$I_{D1} = \frac{g_m V_1}{2\pi}\left[\frac{1}{2}\left(\phi + \frac{\sin 2\phi}{2}\right) - 2\cos\frac{\theta}{2}\sin\phi\right]\Bigg|_{-\frac{\theta}{2}}^{\frac{\theta}{2}} \tag{6}$$

$$I_{D1} = \frac{g_m V_1}{2\pi}(\theta - \sin\theta) \tag{7}$$

Later, the mean-square value of a unit rectangular window spanning the conduction angle θ is used. This is calculated as follows:

$$\overline{w^2} = \frac{1}{2\pi}\int_{-\pi}^{\pi} w^2(\phi)\, d\phi \tag{8}$$

$$\overline{w^2} = \frac{\theta}{2\pi} \tag{9}$$

Turning to circuit equations, the sum of the voltages around the resonator leads to (10).

$$V_{C1} + sC_1 V_{C1}(sL + R) = -\frac{1}{sC_2}(I_D + sC_1 V_{C1}) \tag{10}$$

Collecting terms of V_{C1} and only looking at harmonics near the oscillation frequency leads to (11).

$$V_{C1}\left[1 + \frac{C_1}{C_2} - \omega^2 LC_1 + j\omega RC_1\right] = -\frac{I_{D1}}{j\omega C_2} \tag{11}$$

From the operation of the circuit, it is known that V_1 is in phase with I_{D1}. Therefore (11) is simplified by first equating real parts on both sides.

$$1 + \frac{C_1}{C_2} - \omega^2 L C_1 = 0 \tag{12}$$

This leads to the frequency of oscillation,

$$\omega^2 = \frac{C_1 + C_2}{L C_1 C_2}. \tag{13}$$

Next, equate the imaginary parts.

$$V_1 = \frac{I_{D1}}{\omega^2 C_1 C_2 R} \tag{14}$$

Equations (7) and (13) simplify (14) and specify the relation between g_m and the transistor conduction angle, θ.

$$V_1 = \frac{g_m V_1}{2\pi}(\theta - \sin\theta)\frac{L}{(C_1 + C_2)R}, \tag{15}$$

which leads to

$$\frac{(\theta - \sin\theta)}{2\pi} = \frac{(C_1 + C_2)R}{g_m L}. \tag{16}$$

As expected, when g_m gets very large, θ tends to 0. Finally, using DC balance between the circuit bias current and I_{D0} from (3).

$$V_1 = \frac{2\pi I_0}{g_m\left(2\sin\frac{\theta}{2} - \theta\cos\frac{\theta}{2}\right)} \tag{17}$$

Limiting Case: What is the amplitude of oscillation if the transistor g_m is very large, as in a BJT? V_1 is found by eliminating g_m between (15) and (17):

$$V_1 = \frac{(\theta - \sin\theta)}{\left(2\sin\frac{\theta}{2} - \theta\cos\frac{\theta}{2}\right)}\frac{I_0 L}{(C_1 + C_2)R} \tag{18}$$

Next, the limit as θ tends to 0 is found. Initially this leads to an indeterminate answer. However, by applying L'Hopital's rule successively until the numerator and the denominator are no longer both zero results in:

$$V_1 = \frac{2I_0L}{(C_1+C_2)R} \quad \text{as} \quad \theta \to 0 \tag{19}$$

These results are exactly the same as both equations (14) and (15) in Huang [3].

The interesting thing to note is that the amplitude, and therefore the fundamental component of the transistor current, stays almost constant over a very large range of conduction angle. This is shown in Figure 3, which plots the first part of (18) for various values of θ. Therefore, for a reasonable conduction angle, the Fourier series of I_D is accurately approximated as:

$$I_D \cong I_0 + 2I_0 \cos\omega t + \ldots \tag{20}$$

3 Phase Noise Analysis

While the noise analysis by Huang is accurate, the mathematics involved are so complicated it is difficult to develop insight into the phase noise processes of the Colpitts oscillator [2,3]. The following analysis is simpler than Huang's but leads to exactly the same conclusion.

3.1 Noise Sources

The analysis of the noise in the oscillator begins by identifying the three noise sources in the circuit. First there is the resistor noise, which is represented as a series noise voltage with constant spectral density:

$$\hat{v}_{\mathrm{n}}^2 = 4kT\gamma R \tag{21}$$

Second is the noise in the bias current source, which is represented as a shunt noise current with constant spectral density:

$$\hat{i}_{\mathrm{n,\,cs}}^2 = 4kT\gamma g_{\mathrm{m,\,bias}}, \tag{22}$$

FIGURE 3 *The first term of (18) as a function of conduction angle θ. For narrow conduction angles, this is accurately approximated by 2.*

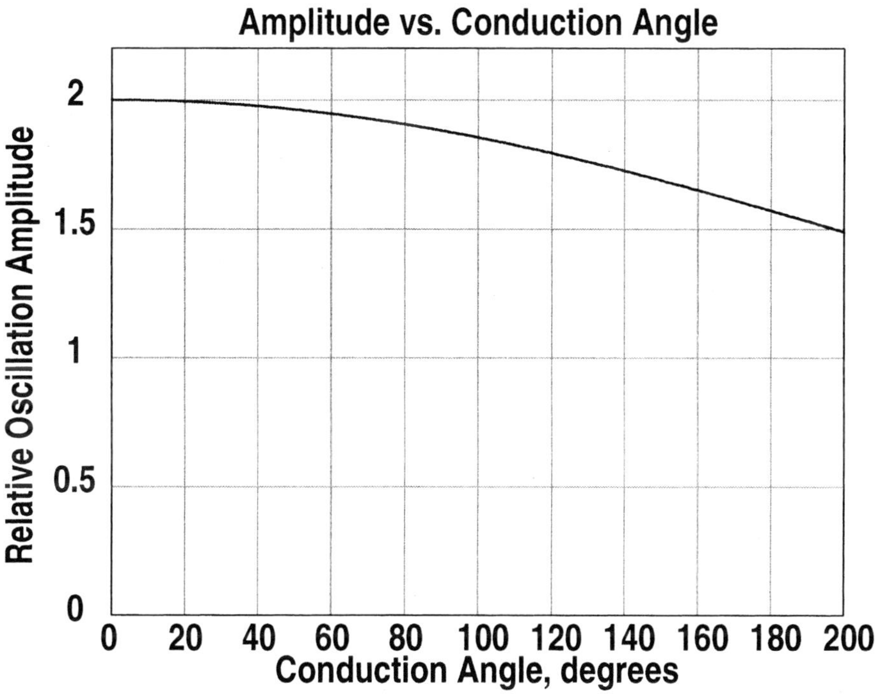

where F_1 is the appropriate noise factor. Finally, there is the noise in the switching transistor. This noise is gated in time; that is, it is zero everywhere except when the transistor turns on for the conduction angle θ. The equivalent continuous spectral density of this noise is derived below.

The output current noise of the transistor is described along the angle axis (equivalently ωt axis) as:

$$i_{n1}(t) = i_{n0}(t)w(t) \tag{23}$$

where the noise of a transistor is $\hat{i}_{n0}^2(f) = 4kT\gamma g_m$ and γ is 2/3 for MOS devices. The window function $w(t)$ was defined earlier. Multiplication in time corresponds to convolution in the frequency domain. Using the fact that the original noise is white, the resulting spectral density is:

$$\hat{i}_{n1}^{2}(f) = \sum_{k=0}^{\infty} [i_{n0}(f) \otimes W_k(f)]^2 = \hat{i}_{n0}^{2} \sum_{k=0}^{\infty} [W_k(f)]^2 \tag{24}$$

In the right-most term, the sum of the squares of all the harmonics is equal to the mean-square value of the periodic window, which can be directly computed in the time domain. This was computed previously in (8).

$$\hat{i}_{n1}^{2}(f) = 4kT\gamma g_m \frac{\theta}{2\pi} \tag{25}$$

This is a constant spectral density representing gated noise and is equivalent to Huang's expression, whose derivation he explains in the text that precedes it. Using (16), this is simplified by expressing g_m in terms of conduction angle and circuit constants,

$$\hat{i}_{n1}^{2}(f) = 4kT\gamma \frac{(C_1 + C_2)R}{L} \frac{\theta}{\theta - \sin\theta}. \tag{26}$$

Limiting case: The question is, what is this spectral density as g_m gets very large and θ tends to zero? Applying L'Hopital's rule, we find that this output current noise spectral density grows without bound. This does not seem credible, but Huang's expression has the same property, as his β approaches infinity (infinitely strong MOSFET) and x tends to 1 (conduction angle goes to zero). However, it must be true since the noise is present for an infinitesimally small interval of time.

Since the current source noise and the noise from the transistor are essentially in parallel and are uncorrellated, they can be added powerwise.

$$\hat{i}_{n}^{2}(f) = \hat{i}_{n1}^{2}(f) + \hat{i}_{n,\,cs}^{2} \tag{27}$$

3.2 Noise in the Resistor

First, the phase noise of the current source is analyzed. The resonator filters all the noise in the current source that does not lie near the frequency of oscillation. Therefore, only noise around the carrier is important. A noise current located at the upper sideband is analyzed.

The first step is to assume an output voltage, (28). This equation is slightly different from the solution that was assumed in Section 3.1 in Chapter 3. It consists of a noiseless carrier plus PM sidebands. It is assumed and AM sidebands are negligible near the carrier. The level of the PM components (represented by *a*) are unknown but the amplitude of the noiseless oscillation is. The amplitude can be calculated from (18).

$$V_{C1} = V_1 e^{j\omega t} + \left(a e^{j\omega_+ t} - a^* e^{j\omega_- t}\right) \tag{28}$$

An output current is also assumed. However because of the highly nonlinear shape of the current, the relative amount of AM is unknown and must be included in the expression. This is shown in (29).

$$I_D = I_{D0} + G_m V_1 e^{j\omega t} + G_m\left(a e^{j\omega_+ t} - a^* e^{j\omega_- t} + b e^{j\omega_+ t} + b^* e^{j\omega_- t}\right) \tag{29}$$

This is simplified by using (20).

$$I_D = I_0 + 2I_0 e^{j\omega t} + \frac{2I_0}{V_1}\left(a e^{j\omega_+ t} - a^* e^{j\omega_- t} + b e^{j\omega_+ t} + b^* e^{j\omega_- t}\right) \tag{30}$$

As was done before in (10), the voltages around the resonator are summed. However, this time the noise of the resistor, defined in (21), is included.

$$V_{C1} - v_n + sC_1 V_{C1}(sL + R) = -\frac{1}{sC_2}(I_D + sC_1 V_{C1}) \tag{31}$$

Once again, the terms in V_{C1} are collected and only the harmonic near the fundamental are kept. Effectively, the noise voltage of the resistor induces a voltage on V_{C1}.

$$V_{C1} = \frac{v_n}{F(j\omega)} - \frac{I_D}{j\omega C_2 F(j\omega)} \tag{32}$$

where

$$F(j\omega) = 1 + \frac{C_1}{C_2} - \omega^2 LC_1 + j\omega C_1 R \tag{33}$$

Assuming that v_n is the amplitude of an injected tone representing the resistor noise, and its frequency is ω_+, the following equation governs harmonic balance at this frequency:

$$a = \frac{v_n}{F(j\omega_+)} - 2I_0\left(\frac{a}{V_1} + \frac{b}{V_1}\right)\frac{1}{j\omega_+ C_2}\frac{1}{F(j\omega_+)} \tag{34}$$

whereas harmonic balance at $\omega\_$ leads to:

$$-a^* = -2I_0\left(\frac{-a^*}{V_1} + \frac{b^*}{V_1}\right)\frac{1}{j\omega\_C_2}\frac{1}{F(j\omega\_)} \tag{35}$$

Taking the conjugate of both sides of this expression:

$$-a = -2I_0\left(\frac{a}{V_1} - \frac{b}{V_1}\right)\frac{1}{j\omega\_C_2}\frac{1}{F(j\omega\_)} \tag{36}$$

This last expression can be used to find the AM sidebands, b, in terms of the PM sidebands, a:

$$\frac{b}{V_1} = \frac{a}{V_1}\left(1 - \frac{L}{R(C_1 + C_2)}j\omega\_C_2 F^*(j\omega\_)\right) \tag{37}$$

Substituting this in (34), leads to:

$$-\frac{v_n}{F(j\omega_+)} = \frac{a}{A}2I_0\left(\frac{L}{R(C_1 + C_2)}\left(1 - \frac{\omega\_F^*(j\omega\_)}{\omega_+ F^*(j\omega_+)}\right) + \frac{2}{j\omega_+ C_2 F(j\omega_+)}\right) \tag{38}$$

This can be rewritten as:

$$-\frac{v_n}{F(j\omega_+)} = \frac{a}{A}2I_0\left(\frac{\omega_+ F(j\omega_+) - \omega\_F^*(j\omega\_) - 2j\omega^2 C_1 R}{\omega^2 C_1 C_2 R\omega_+ F(j\omega_+)}\right) \tag{39}$$

The numerator of the RHS is a small number. As the offset frequency approaches zero, the sum of the first two terms in the numerator are equal and opposite to the third term. Therefore, it is expected to be proportional to the offset frequency, which is exactly the format expected for phase noise sidebands. The following expression is arrived at by ignoring 2nd order terms in (ω_m/ω_0):

$$-v_{\mathrm{n}} = \frac{a}{V_1} 2I_0\left(-\frac{4\omega_{\mathrm{m}}L}{C_2R(\omega+\omega_{\mathrm{m}})}\right), \tag{40}$$

$$\frac{a}{V_1} = v_{\mathrm{n}}\frac{C_2R(\omega+\omega_{\mathrm{m}})}{2I_0\times 4\omega_{\mathrm{m}}L}. \tag{41}$$

Now, rewriting (19), the expression for amplitude of oscillation,

$$V_1 = 2I_0\frac{L}{R(C_1+C_2)} \tag{42}$$

the phase noise sideband strength simplifies to:

$$a = \frac{v_{\mathrm{n}}}{4}\frac{C_2}{C_1+C_2}\frac{\omega_0}{\omega_{\mathrm{m}}} \tag{43}$$

As expected in an oscillator where the restoring current cancels the resistor loss, and where PM passes unchanged through the restoring current source, the phase noise sidebands rise without limit approaching the carrier.

However, the response to AM should be different. Substituting the PM result, (43), into the expression for AM sideband, (37), we get:

$$b = a\left[1 - j\frac{LC_2}{R(C_1+C_2)}(\omega_0-\omega_{\mathrm{m}})\left(1+\frac{C_1}{C_2}-(\omega_0-\omega_{\mathrm{m}})^2LC_1 - (\omega_0-\omega_{\mathrm{m}})C_1R\right)\right], \tag{44}$$

$$b \approx a\left(\frac{\omega_{\mathrm{m}}}{\omega_0} - j\frac{2\omega_{\mathrm{m}}L}{R}\right), \tag{45}$$

which simplifies to:

$$b \approx \frac{v_{\mathrm{n}}}{4}\frac{C_2}{C_1+C_2}(1-2jQ), \tag{46}$$

$$b \approx -j\frac{v_n}{4}(2Q)\left(\frac{C_2}{C_1 + C_2}\right). \tag{47}$$

This shows that, as expected, close to the oscillation frequency, the AM sidebands are constant and depends on the resonator Q, which defines the resonator resistance. There is a slight inconsistency, however. It was assumed that there were no AM sidebands on V_{C1}, whereas the presence of b will induce AM voltage sidebands on C_1.

3.3 Noise of the Current Source and the Transistor

Next, analyze the response of the circuit to current noise in parallel with the current source. The governing equation is:

$$v_{gs} = -\frac{I_M + i_n}{j\omega C_2\left[\left(1 + \frac{C_1}{C_2} - \omega^2 LC_1\right) + j\omega C_1 R\right]} \tag{48}$$

The analysis proceeds identically to that with the noise in the resistor. Indeed, all it requires is the substitution of v_n by $i_n/j\omega C_2$. Thus, the PM on V_{C1} is:

$$a = -\frac{1}{4}\frac{i_n}{j\omega C_2}\frac{C_2}{C_1 + C_2}\left(\frac{\omega_0}{\omega_m}\right) = -\frac{1}{4}\frac{i_n}{j\omega(C_1 + C_2)}\left(\frac{\omega_0}{\omega_m}\right) \tag{49}$$

which is identical to the outcome of Huang's analysis.

3.4 Noise Factor of Colpitts Oscillator

Next, the noise factor of the Colpitts oscillator, as specified by the normalized expression taken from Leeson for phase noise in any oscillator, is determined:

$$\mathcal{L}(\omega_m) = \frac{4FkTR_p}{V_0^2}\left(\frac{\omega}{2Q\omega_m}\right)^2 \tag{50}$$

where $4kTR_p$ is the voltage noise spectral density across the standalone resonator at resonance. The amplitude of oscillation across the resonator is V_0 and Q is the quality factor of the tank. The noise factor F allows comparison of different types of oscillators. Since all the calculations for noise were done with respect to the voltage across the capacitor, they must be scaled.

$$V_{n0} = \frac{C_1 + C_2}{C_2} V_{nC_1} \tag{51}$$

The total PM white noise voltage spectral density is found by squaring and combining (43) and (49). Next this sum is scaled by the factor in (51):

$$\hat{v}_{n0}^2(f) = \left(\frac{C_1 + C_2}{C_2}\right)^2 \left[\frac{\hat{v}_n^2}{16}\left(\frac{C_2}{C_1 + C_2}\right)^2 + \frac{\hat{i}_n^{\,2}}{16} \frac{1}{\omega^2 (C_1 + C_2)^2}\right] \left(\frac{\omega}{\omega_m}\right)^2 \tag{52}$$

After some simplification:

$$\hat{v}_{n0}^2 = \left(\frac{\hat{v}_n^{\,2}}{16} + \frac{\hat{i}_n^{\,2}}{16} \frac{1}{\omega^2 C_2^2}\right) \left(\frac{\omega}{\omega_n}\right)^2 \tag{53}$$

Finally, expand the noise current to include the noise from the current source and the switching device using (27).

$$\hat{v}_{n0}^2 = \frac{kT}{4}\left\{R + \frac{1}{\omega^2 C_2^2}\left(\gamma \frac{(C_1 + C_2)R}{L} f(\theta) + \gamma g_{m,\text{bias}}\right)\right\}\left(\frac{\omega}{\omega_m}\right)^2 \tag{54}$$

where

$$f(\theta) \approx \frac{\theta}{\theta - \sin(\theta)}. \tag{55}$$

Taking into account that while noise above and below the oscillation frequency will contribute to phase noise at a given offset, the spectral density of phase noise is:

$$\mathcal{L}(\omega_m) = \frac{2 \hat{v}_{n0}^2}{\frac{1}{2} V_1^2} = \frac{kT}{V_1^2}\left\{R + \frac{1}{\omega^2 C_2^2}\left(\gamma \frac{(C_1 + C_2)R}{L} f(\theta) + \gamma g_{m\ \text{bias}}\right)\right\}\left(\frac{\omega_0}{\omega_m}\right)^2 \tag{56}$$

$$\mathcal{L}(\omega_m) = \frac{4kTR_p}{V_1^2}\left[1 + \gamma\frac{C_1}{C_2}f(\theta) + \gamma g_{m\,\mathrm{bias}}\frac{C_1}{C_2}\frac{L}{(C_1 + C_2)R}\right]\left(\frac{\omega_0}{2Q\omega_m}\right)^2 \quad (57)$$

where $R_p = Q^2 R$.

Compare with (57) from Chapter 3 for the phase noise in a differential LC oscillator:

$$\mathcal{L}(\omega_m) = \frac{4kTR}{V_1^2}\left[1 + \frac{2\gamma I_0 R}{\pi V_1} + \frac{\gamma I_0 R}{2V_{\mathrm{eff}}}\right]\left(\frac{\omega_0}{2Q\omega_m}\right)^2 \quad (58)$$

The noise factor expressions are remarkably alike for both oscillators, except for the capacitor tapping factor C_1/C_2 that scales the 2nd and 3rd terms in the sum. Also, R is the equivalent series resistance representing tank loss, and R_p the shunt resistor.

If anything, as $f(\theta)$ appears to grow without bound, the Colpitts oscillator will be noisier than the differential LC oscillator at narrow conduction angles! This would be true even if we filter white noise in the bias current using Hegazi's technique [4].

4 Conclusions

The Colpitts oscillator does not appear to be an inherently better topology than the differential pair oscillator. For example, some of the current supplied by the transistor to overcome the loss in the resonator flows through C_2 making this topology less efficient than the differential LC oscillator. This should result in a lower figure of merit. However, even though the Colpitts does not offer inherently better phase than the current biased oscillator, it is still a useful circuit in application where exceptional phase noise is required dictating a high Q off-chip single-pin resonator.

References

[1] E. V. Kulagin, A. I. Pikhtelev, V. P. Sokolov, and B. P. Fateev, "Natural fluctuations in a quartz crystal oscillator with automatic gain control," *Radiophysics and Quantum Electronics*, vol. 21, pp. 1125-30, 1978.

[2] Q. Huang, "On the exact design of RF oscillators," *Proceedings of the IEEE 1998 Custom Integrated Circuits Conference*, 1998.

[3] Q. Huang, "Phase noise to carrier ratio in LC oscillators," *IEEE Transactions on Circuits and Systems I: Fundamental Theory and Applications*, vol. 47, no. 7, pp. 965-980, Jul. 2000.

[4] E. Hegazi, H. Sjoland, and A. A. Abidi, "A filtering technique to lower LC oscillator phase noise," *IEEE Journal of Solid-State Circuits*, vol. 36, no. 12, pp. 1921-1930, Dec. 2001.

5 Design for Low Thermal Phase Noise

1 Introduction

A single chip radio remains a challenging problem due to technology limitations on passive component quality and lack of efficient optimization procedures. Of all RF blocks, voltage controlled oscillators, VCOs, have received highest attention in recent years as evidenced by the large number of publications reporting improving performance [1][2], higher operating frequency [3] or using a different passives technology to achieve the stringent requirements of wireless standards [4]. Integrated LC oscillator circuits published so far use tuning inductors that are fully integrated, partly integrated, or discrete, with quality factors spanning a large range. However, lacking a clear understanding of the physical processes of phase noise, it is difficult to compare the relative merits of these VCOs in a normalized sense. In this chapter, we present a non-exhaustive set of differential CMOS LC oscillators illustrating a systematic design methodology that builds on the phase noise theory presented earlier.

1.1 Oscillator Figure of Merit

The design space of LC oscillators entails phase noise, power consumption and oscillation frequency, and to a lesser degree, tuning range. These design dimensions are not always orthogonal and a well-formed cost function is necessary for comparing the relative merits of various designs. The best definition of a normalized Figure Of Merit (FOM) proposed so far is [5]:

$$\mathrm{FOM} = \left(\frac{\omega_o}{\omega}\right)^2 \frac{1}{\mathcal{L}(\omega)P} \tag{1}$$

where P is the power consumption in mW, $\mathcal{L}(\omega)$ is the phase noise at an offset ω from a center frequency ω_0.

The appeal of this definition is that it relates and normalizes the quantities given by Leeson's proportionality. The power of two in the frequency offset dependence indicates that flicker noise-induced phase noise is not normalized, suggesting that the FOM is useful only in comparing oscillators at large-enough offsets such that flicker noise is not the dominant source of phase noise. This chapter focuses on minimizing thermal noise's impact on oscillator phase noise. The FOM described above will be used as the inverse cost function to be maximized.

2 Note About Harmonic Balance in LC Oscillators

The LC tank, assuming no losses, is a typical physics textbook oscillator. Magnetic energy in the inductor converts to electric energy in the capacitor and vice versa. The oscillation is governed by the energy equation, which states that the total energy stored on the capacitor and in the inductor is constant.

The energy balance equation is not sufficient to describe the oscillation. The reason for that is the need to know the total oscillator energy, which is set by the initial current in the inductor and the initial voltage on the capacitor.

In practical oscillators, there are losses that must be overcome if the oscillation is to continue forever. In LC oscillators, the losses are overcome by employing a negative resistance, which in practice is implemented with one or more nonlinear components. They are represented in Figure 1 as the "non-sinusoidal" current source. The negative resistance current enters the tank and builds an oscillation amplitude that equals the product of the tank resistance at resonance and the fundamental of the current waveform.

$$A = RI_{\text{fund}} \tag{2}$$

The capacitor offers little impedance to the harmonics of the injected current flow. On the other hand, the inductor impedes the harmonics from passing through it. The fundamental of the injected current flows only through the resistor since the inductor and the capacitor represent an open circuit at resonance. Only the reactive current can flow through the tank. The power in the inductor has to balance with that in the capacitor. However, the harmonics result in a power imbalance since they flow in unequal amounts through the capacitor and the inductor. How does the oscillator handle the situation? By forcing the power in the harmonics to be smaller. This is achieved by shifting the oscillation frequency down [6]. Therefore, in the presence of the harmonics of the negative resistance current, the tank is always oscillating below its natural frequency. The tank appears more inductive and the power in the fundamental tone through the inductor is higher than that through the capacitor. This excess of fundamental's power, balances the power in the harmonics that flow through the capacitor. We will revisit this result later when we analyze flicker noise.

FIGURE 1 *Current split in an LC oscillator.*

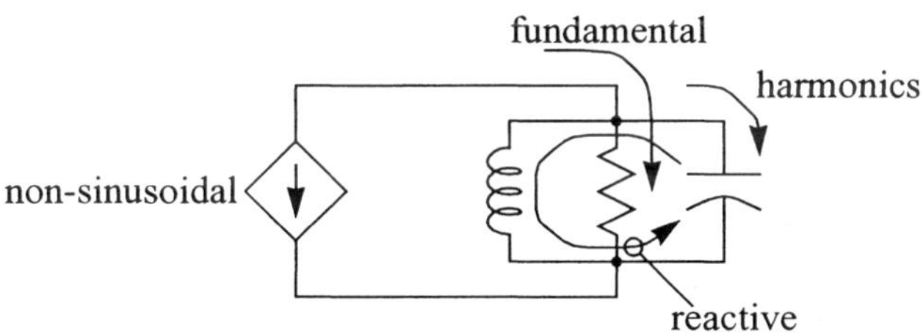

3 Amplitude in Differential LC Oscillators

3.1 Current-Biased Differential Oscillators

A popular implementation of an LC oscillator is the current-biased negative-g_m topology shown in Figure 2. The reasons for its popularity are that it provides differential and often quadrature differential outputs in addition to its well-controlled power consumption and its easy-to-guarantee start-up, which makes it a "safe" topology. In the oscillators shown in the figure, the resonator capacitors take the form of M_3 and M_4, which are acting as varactors, or voltage controlled capacitors.

The differential negative-g_m oscillator comes in different variants. First, it is either voltage biased or current biased. The voltage-biased VCO is similar to those shown in Figure 2 with the current source device M_0 replaced by a short

FIGURE 2 *Current-biased differential LC oscillator: tail and top-biased.*

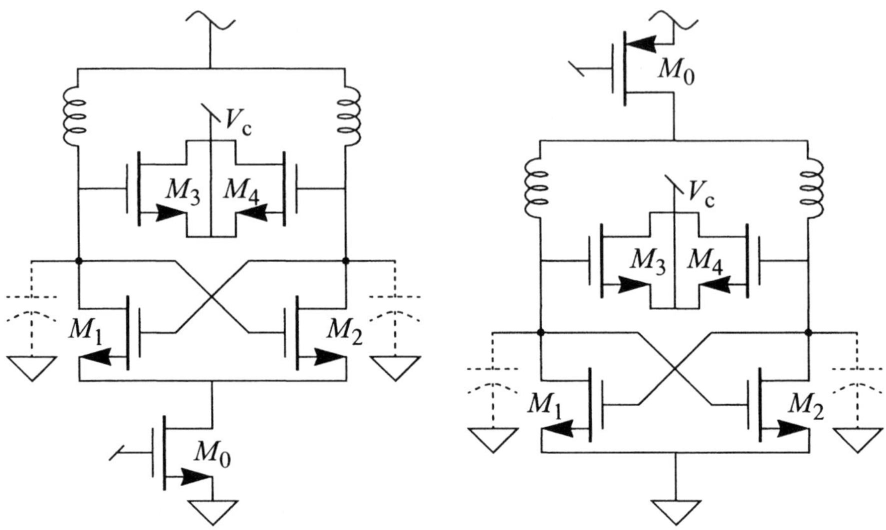

circuit to ground. The current-biased oscillator comes in various forms: top-biased, tail-biased and complementary-differential. The top-biased and the tail biased topologies, shown in Figure 2, seem isomorphic at first glance with the difference being the reference point, which is supposedly arbitrary. In other words, if we consider the VCO core to be the LC tank and the cross coupled differential pair providing the negative resistance, feeding the bias current from the top or from the bottom seems like a minor detail. However, when we look closer at the grounded parasitic capacitors and the current paths within the oscillator, differences between the two topologies become clear. These differences can lead to different trade-offs in the design procedure as we will illustrate later.

Consider the cross-coupled differential pair formed by M_1 and M_2 in Figure 2. It provides a small signal negative differential conductance of $-g_m$ across the tuned circuit. When properly designed, the negative conductance overcomes the positive loss conductance of the LC tank. The natural response of this two pole circuit is a growing oscillation that is eventually limited in amplitude by circuit nonlinearity. When the differential pair is biased at low currents, this nonlinearity stems from bias current exhaustion. When the current is fully switched from one side to the other, the current in one device is doubled from that at the quiescent condition. Suppose the oscillation amplitude is larger than the voltage required to commutate the differential pair current, $\sqrt{2}(V_{GS} - V_T)$, the differential pair sustains the oscillation by injecting an energy-replenishing square wave current into the LC resonator. The second order bandpass LC resonator rejects the harmonics and responds to the fundamental frequency in the current waveform with a single-ended voltage amplitude of oscillation:

$$V_o = \eta I R \tag{3}$$

In that the amplitude of oscillation is proportional to the bias current, this is known as the *current-limited* regime. Note that R is the single-ended tank resistance.

$$R = Q\omega L \tag{4}$$

We will define the ratio between the single-ended fundamental current (that flows through the tank resistor at f_0) and the bias current to be, the current switching efficiency η:

$$\eta \equiv \frac{I_{fund}}{I_o} \tag{5}$$

In the current-biased oscillator shown, the current injected into the tank is ideally a square wave between I_0 and 0 leading to a current efficiency of $2/\pi$. In real designs, the switching efficiency is closer to 0.5 than $2/\pi$. Since the current is switched between the two sides of the tank, the differential swing is twice that given by (3). As the tail current is raised in value, the amplitude also rises until, approaching a single-ended amplitude of almost V_{DD}, negative peaks momentarily force the current source transistor into triode region. This is a self-limiting process. Raising the gate bias on the current source forces this FET to spend a greater fraction of the oscillation cycle in triode, resulting in no appreciable rise in the amplitude. This is called the *voltage-limited* regime, whose onset is defined by entry of the current source transistor into triode region. Further increase of the gate source voltage of the current source will lead to smaller growth of oscillation amplitude as the current source device is pushed more and more into triode operation and the common mode point is pulled closer to ground. In the limiting case, if we assume that the gate oxide of the current source device can withstand an infinite voltage, the common mode point is at zero volts and the oscillator is identical to a voltage biased oscillator. The former description suggests that the oscillator core has a particular maximum current that cannot be exceeded. The oscillator core "resists" the increase of the current by pushing the current source into triode more and more. The maximum current that can be fed into the core is, therefore, equal to that of the voltage-biased VCO, which sets the VCO at maximum amplitude equal to the supply voltage. That is:

$$I_{dc} = \frac{V_{DD}}{\eta Q_{loaded}\omega L} \tag{6}$$

where Q_{loaded} is the loaded quality factor or the resonator. The amplitude dependence on the bias current is shown in Figure 3 where the amplitude grows linearly with current and once the current source enters triode, the amplitude growth with current slows down significantly but the differential amplitude keeps on growing until it reaches a maximum value of 2 V_{DD} when the current consumption is at its maximum value given by (6).

3.2 Voltage-Biased Oscillator

In a voltage-biased oscillator, Figure 4, the sum of the currents in the two transistors, M_1 and M_2, is not fixed. In fact the circuit is not truly differential but rather *pseudo-differential.* When one transistor is switched on, it enters triode region deeply. The triode FET shorts one side of the resonator to ground with a low output resistance. The oscillation amplitude is limited finally by the power supply. The single ended oscillation voltage sweeps from 0 to 2 V_{DD}. For most of the time, half the tank impedance is shorted. Using simulations the single-ended current switching efficiency is found to be around 0.35 assuming that the effective tank impedance has not been degraded by the triode switches. Alternatively, we can elect to assume the current switching efficiency of $2/\pi$ ($4/\pi$ differentially). In other words, we ascribe the drop in the fundamental current to a lower "loaded" quality factor rather than switching efficiency. This leads to the following approximation

$$Q_{loaded} \approx \frac{Q}{2}, \tag{7}$$

because in a voltage-biased oscillator the effective tank impedance is halved because of the triode FET loading. The conclusion is that voltage-biased

FIGURE 3 *Current-biased oscillator modes of operation*

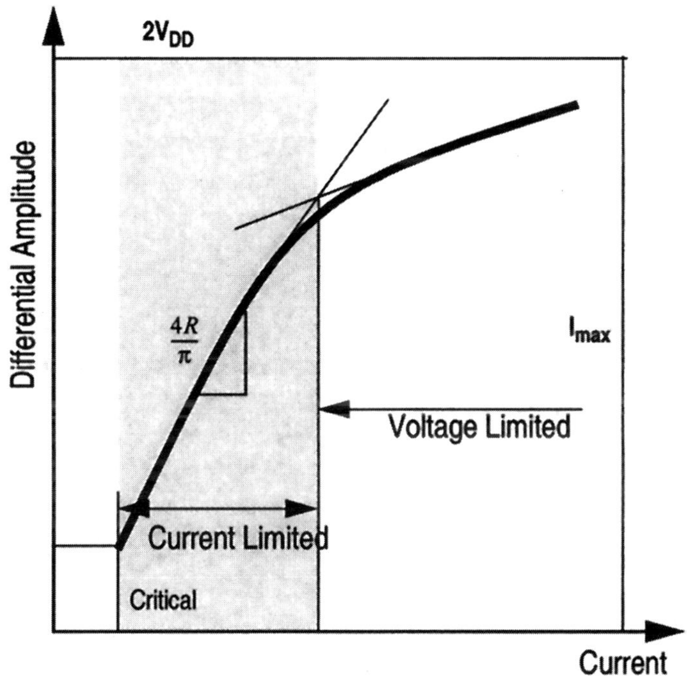

oscillators are less efficient in utilizing bias current to build-up swing. Therefore, they inherently have a lower figure of merit. Almost 6 dB lower than a current-biased oscillator with the same *unloaded* tank quality.

3.3 Colpitts Oscillators

Wang describes the problem of current switching efficiency in a general oscillator [7]. He concludes that, in the domain of real waveforms, the most efficient current waveform is the impulse. This is realized in a Colpitts oscillator (Figure 5) where the transistor conducts for a very short duration (small conduction angle) resulting in an impulse of negative resistance current. The current switching efficiency of the Colpitts oscillator is $\eta_{impulse} = 1$. The resulting amplitude of oscillation is given by:

$$A = I_0 Q \omega L \quad (8)$$

FIGURE 4 *Voltage-biased oscillator.*

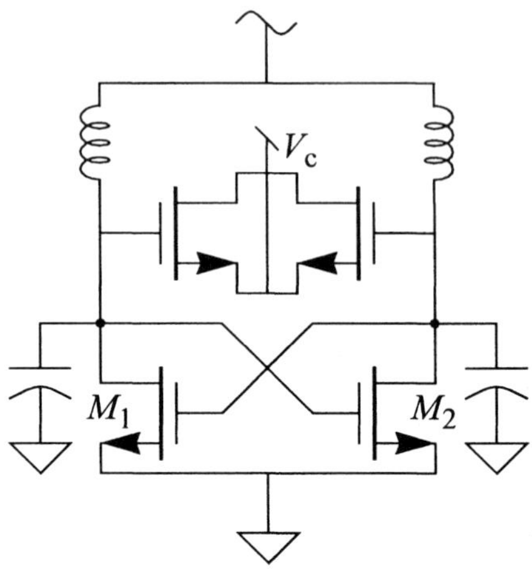

FIGURE 5 *Single-ended Colpitts oscillator.*

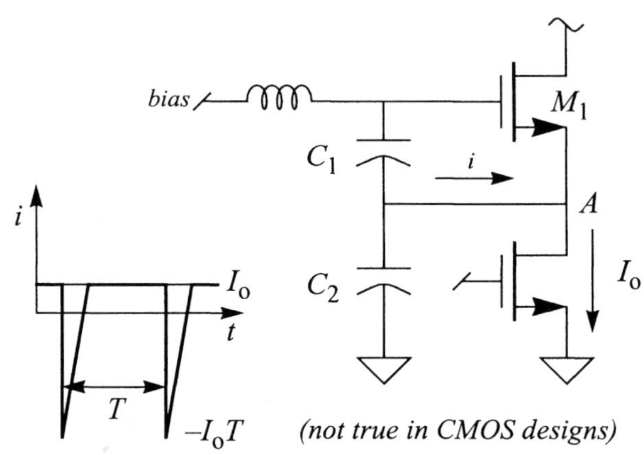

In a bipolar transistor implementation, the current waveform is indeed impulsive thanks to the high transconductance of BJTs. The current switching efficiency is very close to unity. Unfortunately, the current switching waveform in a CMOS implementation is far from impulsive due to the low transconductance of FETs compared to BJT. This lowers the current switching efficiency. Furthermore, the transistor can go into triode (or saturation in a BJT implementation) at a fairly small amplitude, further limiting the amplitude growth.

The typical design strategy thus entails limiting the swing to avoid pushing the transistor into triode. Therefore, it cannot yield good phase noise with the low-quality inductors available in monolithic implementations. This has kept the fully integrated Colpitts implementation from achieving high FOM. Furthermore, CMOS Colpitts oscillators are hard to start-up because they require a large transconductance. For all of these reasons, a Colpitts oscillator is not suitable for CMOS implementation.

It is worthwhile, however, to analyze the amplitude build up in a Colpitts oscillator. When M_1 is off, the current source charges C_1 until the gate-source voltage of M_1 exceeds V_T, at which time M_1 conducts and resulting current rushes through capacitors C_1 and C_2. This recharges C_2 and shuts off the transistor. The resulting current is a train of narrow current pulses, in the limiting case, impulses that are T seconds apart, where T is the period of oscillation. The current into point A is a constant positive current for the entire oscillation cycle except when during the narrow sliver of time where M_1 conducts as shown in (Figure 5). The average current must equal zero because the capacitors do not allow DC current. Therefore, the impulse intensity is equal to I_0T. Performing Fourier transform on the impulse train to get the fundamental current results in:

$$I_{fund} = 2I_o \tag{9}$$

Note that this implementation is single-ended. In practice, the conduction angle of M_1 is not zero. Therefore, the current switching efficiency is lower than 2. In BJT designs, η is close to 1.5 whereas in CMOS designs, η is close to 0.9! A detailed analysis of the amplitude limiting in Colpitts oscillator due to the triode operation of M_1 was performed by Mayaram [8].

A pseudo-differential version of the Colpitts oscillator is also possible by building a mirror oscillator and joining C_2 with its counterpart in the mirror circuit (Figure 6). This topology is rarely used in CMOS implementation as it does not offer any solution to the swing limiting problem neither does it improve the current switching efficiency since it consumes double the bias current to build double the swing. Again, the current switching efficiency is close to 1.5 in a BJT implementation and close to 0.8-0.9 in a CMOS implementation with a limited swing.

FIGURE 6 *Pseudo-differential Colpitts oscillator.*

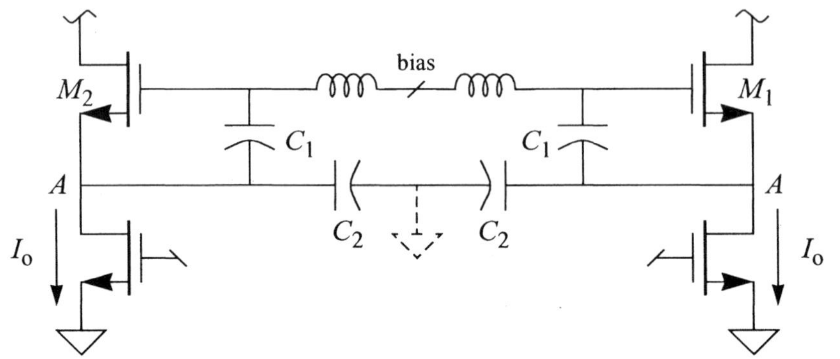

Can the two current sources in the pseudo-differential Colpitts be commutated to boost the current switching efficiency to 2 [9]? To answer this question we need to go back to the amplitude build-up process in the Colpitts oscillator. The negative resistance current strikes the tank for a very short time, almost like an impulse. For the rest of the cycle, the current is constant and equal to the bias value. If a single current source is commutated (Figure 7), the current into the capacitors still needs to have zero average. Assume that the current is fully switched into one side. Transistor M_1 cannot carry the current until a large enough voltage has developed on C_1. Therefore the voltage at node A will keep on dropping linearly in a relaxation process until enough voltage develops on C_1 and momentarily, M_1 turns on for a very short time. An impulsive current passes through M_1 into the capacitors and shuts M_1 off. This happens when the voltage at A is at minimum. Therefore this is the moment where the switching transistors flip polarity. The relative phases of this idealized model are depicted in Figure 8. The intensity of the impulsive current is set by the zero average condition. That is the net current passing through the capacitor per cycle is zero. Since a current I passes for a time $T/2$, the impulse current is given by:

$$i_{M1}(t) = -\frac{I_o T}{2}\delta(t) \tag{10}$$

As can be seen, the switched current by M_3 and M_4 is 90 degrees away from that of M_1 and M_2. The amplitude of oscillation is set by the fundamental current passing through the capacitors. Since the two currents in M_1 and M_3 are orthogonal, a vector sum of the currents has to be calculated. In practice however, the impulsive current is not a train delta-dirac impulses (particularly if M_1, M_2 are FETs). In addition, the limited swing in this topology makes the switching of M_3 and M_4 rather soft. In a practical implementation, the phase difference is between 45 and 60 degrees, which makes the vector sum even smaller. Therefore, this circuit never reaches a switching efficiency of 2 (even in a BJT implementation of M_1 and M_2). In addition, like all Colpitts oscillator the circuit suffers from limited swing and amplitude degradation due to triode operation of M_1 and M_2 [8]. A more realistic value for η in this topology is 0.6 in a full CMOS implementation. This is very close to what a differential current-biased VCO can achieve.

FIGURE 7 *Current-commutated differential Colpitts.*

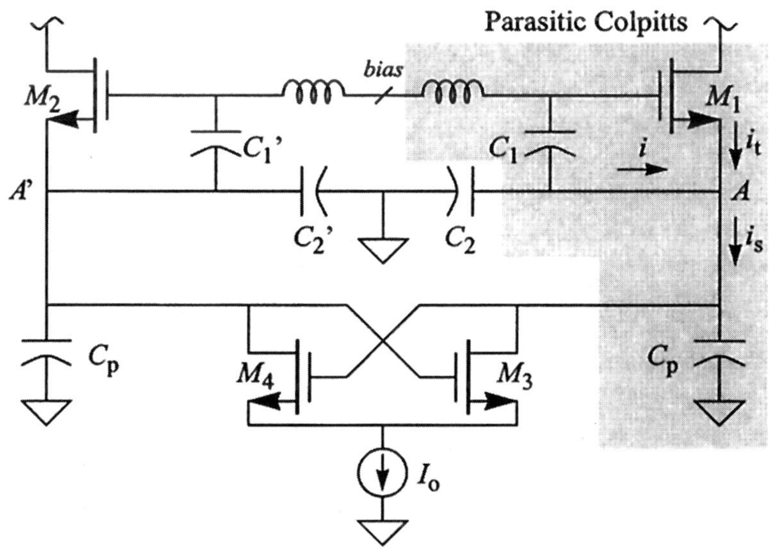

It should be noted that the common point joining C_1 and C_1' should not be left floating if points A and A' have appreciable parasitic capacitance. If this node is left floating in the presence of a large parasitic capacitance at node A, the oscillator has a potential for common-mode oscillation. Recall that this topology is formed using two separate Colpitts oscillators *glued* back-to-back. Each half of the circuit has the potential of oscillating on its own. If points A and A' have a sizable parasitic capacitance, the parasitic can play the role of

FIGURE 8 *Current waveforms in a commutated Colpitts oscillator.*

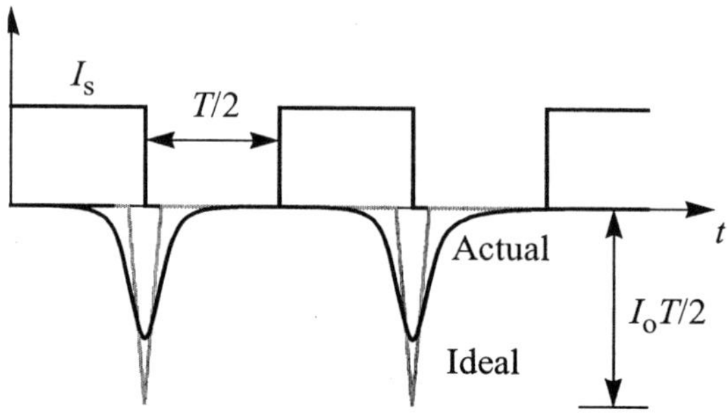

C_1 and initiates a separate oscillation on one half of the circuit. Note that the presence of the cross-coupled switches cannot help because in the case of a common mode oscillation, M_3 and M_4 can go into triode together and maintain an equal split of the bias current at all times. Grounding the common point between C_1 and C_1' lowers the gain for each individual side formed by M_1 and C_1, C_2, and L. Then the only oscillation mode left is the differential one. However, grounding this point lowers the current switching efficiency to a value close to that of the current-biased VCO.

3.4 Complementary Differential Oscillator

Current reuse was extensively utilized in amplifiers and later oscillators to save power consumption. One such circuit is often called complementary differential oscillator [10].

The idea of a current-reuse oscillator is simple and interesting. Since the current is fully-switched from one side to the other then it makes sense to provide that current through a commutator switch operating on a current source of half that required to for a regular current biased oscillator. As shown in Figure 9 the current is commutated through a cross-coupled differential pair and driven into a single inductor of twice the value of that in a normal oscillator. In other words, the center tap of the two inductors in the tank is left floating instead of putting it to a fixed potential. For the same bias current, this topology offers twice the differential swing of that in a current biased oscillator. This is because the current switching efficiency is doubled. However, now there are more devices to contribute noise. The topology promises a 3 dB improvement in FOM over the topologies in Figure 2. In practice, an improvement of around 2 dB is achievable with this arrangement. the reason for that has to do with the extra parasitic capacitance to ground incurred due to the extra crossed coupled pair. Particularly since PMOS FETs have larger junction capacitance per unit of transconductance. Later, we will show that the peak FOM achievable using this topology is the same as that of a regular tail-biased oscillator. The difference is that the complementary differential circuit reaches its peak FOM at a lower current consumption. i.e. at lower output signal swing.

Let's consider two cases: first assume an off-chip inductor, in this case there

FIGURE 9 *Complementary differential oscillator.*

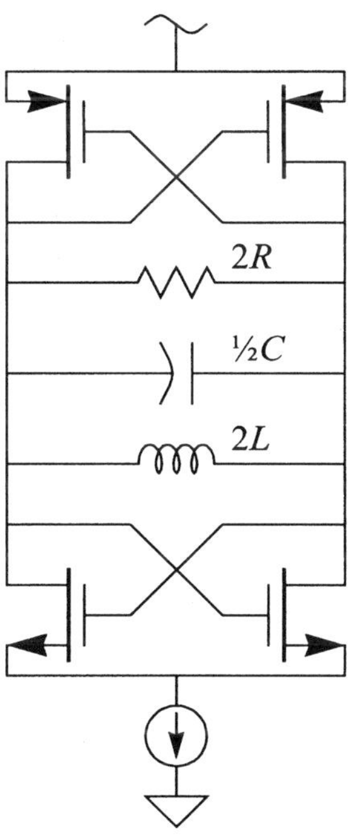

is a large fixed capacitance due to the pad and package. The additional capacitance is not the limiting factor for the inductor value because it is not dominant. In the second case, we consider either a bondwire or an on chip spiral inductor to be used in the tank. The original fixed capacitance is not very large and to maintain a particular tuning range, the tank inductance must be reduced and the current is increased to maintain the swing. Therefore, the performance advantage of this topology is for oscillators with off-chip inductors. There are other reasons that favor the complementary implementation that will be discussed in Section 11. Applying the noise filter technique described in Section 10 yields the best performance possible from this topology.

Many issues regarding the flicker noise up-conversion in the complementary differential topology have been reported in the literature [11,12]. Based on time domain waveforms, it is believed that if the pull-up and pull-down switching times of the oscillator are matched then 1/f noise up-conversion is eliminated [13]. In addition, if the oscillator is laid out in a perfectly symmetric manner, common mode signals are eliminated, leading to lower flicker noise up-conversion. However, the more recent analysis of flicker noise mechanisms shows that most up-conversion mechanisms are due to frequency modulation processes rather than mixing processes. In fact, mixing low frequency noise, predominantly flicker, can only yield amplitude modulation because it produces two side bands around the carrier with their phases arranged as such of AM noise. In Section 10, a practical way of eliminating flicker noise is shown while the oscillator layout is not necessarily fully symmetric [14].

4 Design of Current-Biased Differential Oscillators

In this section we will develop a design methodology and an optimization procedure that makes use of the physical model of phase noise processes developed in the previous chapter.

The heart of the oscillator is its LC tank. For practical and technological considerations, the inductor is the determinative element in the oscillator. In fully integrated oscillators the low nominal quality factor of on-chip inductors is the limiting factor to high performance. Quality factors of on-chip fixed and voltage controlled capacitors tend to be much higher. On-chip inductors are difficult to design and optimize, therefore, the oscillator designer typically needs to pick an inductor out of a library of inductors, much like discrete bipolar circuit design. The appeal of on-chip inductors, despite their limited and often poor quality, is because they have less parasitics compared to the off-chip or bondwire inductors since the overhead of bonding pad and package parasitics is not present. They also save the cost and board space of off-chip inductor. With this in mind, we will explain how to design and optimize an LC oscillator assuming that the designer has to choose an inductor from a predefined set rather than precisely calculating the required inductance for a particular performance measure. The approach we follow here is useful in quickly-limiting the design space to a particular range of inductance values that meet the required performance. Later, using the same set of equations, a more accurate calculation of the required inductance value is possible even though we don't see much practical value in this calculation without the ability to continuously sweep the inductance.

Let's start again with Leeson's proportionality, only this time, we cast it in a different form in terms of the tank capacitance.

$$\mathcal{L}(f) = \frac{1}{V_o^2}\frac{kT}{C}\frac{\omega_o}{Q}\frac{2}{\omega^2}F \tag{11}$$

The conversion from Leeson's equation, to the form in (11) is done using parallel LC model of the tank and assuming the oscillation frequency to be equal to the resonance frequency of the LC tank. The purpose of the reformulation is to arrive at an expression of phase noise that is free of the inductance value for the reasons illustrated in the preceding paragraph. The inductance is implicitly present in (11) since the quality factor of the inductor is a function of L. Here, phase noise is given as kT/C noise that is shaped in frequency by the LC tank and normalized to the power in the oscillation amplitude. Phase noise is further scaled by a circuit specific noise factor, F, the constant of proportionality that comprises noise contributions from various circuit elements. Being circuit specific, the noise factor, needs to be identified for each oscillator topology in terms of device sizes, current, and other circuit parameters.

The noise factor of an LC oscillator is analogous to the noise factor of any other RF circuit. The difference, however, is that in other RF circuits the noise factor is the total noise of the circuit normalized to the noise of the input characteristic impedance, typically 50 Ω, whereas in LC oscillators, the noise factor is equal to the total oscillator phase noise normalized to phase noise due to the resonator loss, which is the tank's characteristic impedance.

An ideal LC oscillator is composed of an inductor and a capacitor. The noise factor of such a circuit is equal to one if the tank is lossless while the oscillator has no phase noise since the tank quality is infinite. Since practical LC tanks are lossy, a means of providing negative resistance is required to sustain the oscillation. Consider the oscillator shown in Figure 10a where the resonator has a finite loss resistance. An ideal negative resistance compensates the sine-wave signal current through the tank loss resistor. The noise of this ideal negative resistance is equal but uncorrelated to the noise in the resonator loss. This gives rise to a noise factor of 2, the sum of the unity noise factors of both resistors. In practice, a nonlinear active circuit as shown in Figure 10b provides the negative resistance. The average negative resistance over a full cycle is equal to the tank loss even though it might well be operative only over small portions of the oscillation cycle. Assuming the noise factor of the nonlinear active circuit to be γ, the noise factor of the whole oscillator becomes $1+\gamma$. Note that this is the fundamental minimum noise factor for a negative-resistance LC oscillator. As we will show later, bias circuits significantly add to this noise factor.

What (11) reveals is that doubling the tank capacitance while keeping the oscillation frequency and amplitude constant, gives a 3 dB reduction in phase noise. To retain the oscillation frequency, the inductance value needs to be halved. To keep the oscillation amplitude constant, the bias current needs to be doubled as the tank impedance is halved. This is, of course, assuming the tank quality to be independent of the tank inductance in place. The physical interpretation of this result is as follows: an LC tank with half the inductance and double the capacitance will have half the characteristic impedance $R = Q\omega L$. The noise current associated with tank is increased by $\sqrt{2}$, as noise power is doubled. Since the impedance is halved, the noise voltage at the oscillator output is reduced by a factor of $\sqrt{2}$ or 3 dB. This is exactly the resulting improvement in phase noise if the oscillation amplitude is kept constant by doubling the current.

The design methodology we will follow stems from this simple remark. Let's first ignore the noise factor F and consider it independent of the oscillation current. This incorrect approximation is made to reach a first glimpse of the design. Now for particular phase noise specifications, we set a power budget, a maximum current that cannot be exceeded and a particular supply voltage. We choose the smallest inductor possible to result in a particular maximum swing at the oscillator output for the specified current. This means that we select the inductor that has the largest impedance at the frequency of oscillation. Any larger inductor will need a smaller capacitor to tune with it at the

FIGURE 10 *Basic LC oscillator: (a) ideal negative resistance, and (b) nonlinear negative resistance.*

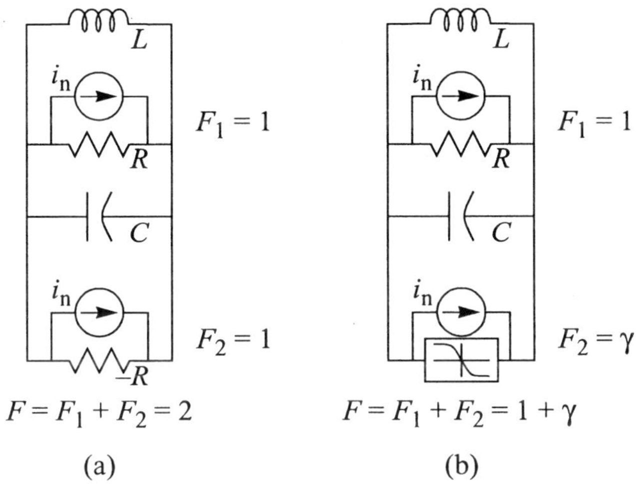

oscillation frequency leading to higher phase noise. To finish one design iteration, the resulting phase noise needs to be compared against the required phase noise specifications. If phase noise is not satisfactory, then the allocated power budget is needs to be increased.

5 A Design Example

In this section we will go through the design phases of a tail-biased oscillator in CMOS. The discussion is carried out from the first steps in design all the way towards layout and measurement results.

5.1 Design Requirements

GSM has phase noise requirements shown in Figure 11. From this plot, it appears that phase noise requirements for low frequency offsets, like 200 kHz, are fairly relaxed. On the other hand, at high frequency offsets, the requirements are stringent. The GSM receive band is 25 MHz centered at 945 MHz. We will assume a high-side injection receiver with LO frequency around 1100 MHz. A tuning range of at least 10% is required to cover for process variations. Target phase noise at 3 MHz offset is below –142 dBc/Hz and maximum current consumption is 3.4 mA from 2.5 V supply.

FIGURE 11 *GSM phase noise requirements.*

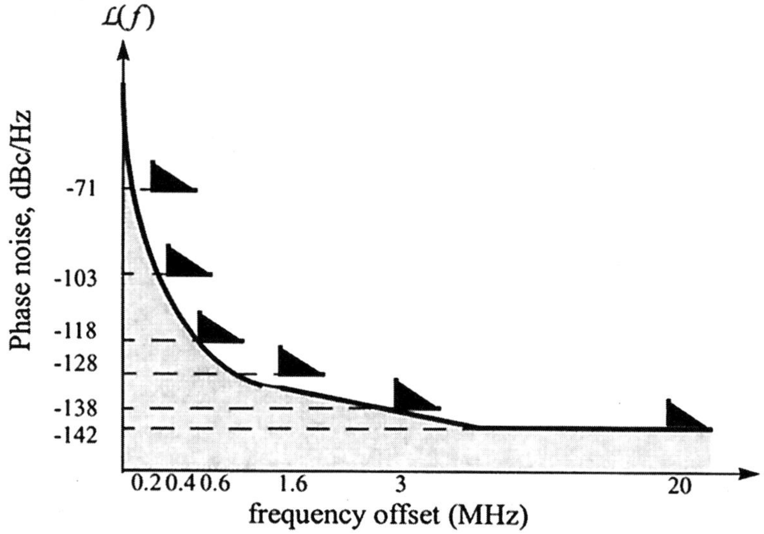

In addition, an inductor library of widespread values and different qualities is provided. A chart of the inductor library values and qualities is given in Figure 12. Using the phase noise expression in (11) and assuming the overdrive voltage of the bias transistor to be 0.4 V, γ_{bias} to be 2/3 and γ to be 4/3, the achieved phase noise is calculated. The distinction between γ and γ_{bias} in value is because for the bias device it is possible to use longer-than-feature channels bringing the channel noise coefficient down to its minimum [15].

The resulting phase noise goes to a minimum close to a tank inductance of around 15-17 nH. In addition, it is necessary to check the startup condition of the oscillator assuming a minimum loop gain of 3. If the overdrive at startup of the switching devices is assumed to be 1 V then a minimum inductance of 8-10 nH is necessary to supply enough impedance to secure startup with a bias current of 3.4 mA. The reader can easily check this claim by calculating the transconductance of the switches and then find the tank impedance required from the loop gain equation.

FIGURE 12 *Sample inductor library.*

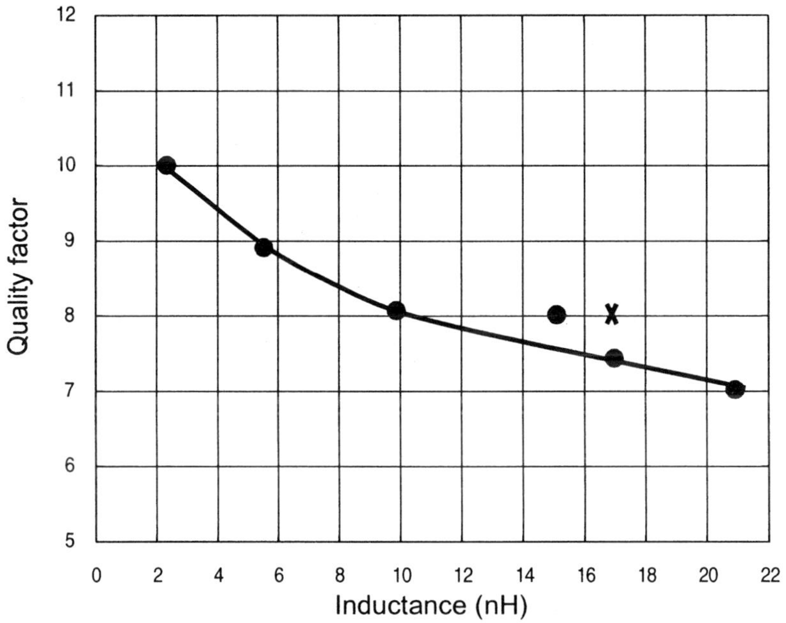

With more optimization of the tank inductor [16], the quality factor of the inductor can be further increased to about 8.4 for a 17 nH inductor, resulting in –142 dBc/Hz at 3 MHz offset. To improve the inductor quality factor, the spiral was altered from a rectangular to an octagonal shape, yielding smaller footprint for a particular inductance. A patterned polysilicon ground shield is also added to reduce displacement current loss [17]. The single-stack metal-4 spiral is also connected in parallel with a similar spiral drawn on metal-3 to reduce the series resistance. Combined, an increase of 15% in Q is achieved.

To make sure that this design is optimal, a plot of the FOM is given in Figure 13. Calculated at 3 MHz offset (thermal noise dominated), the FOM plot shows that small-value inductors result in lower FOM at low currents because of the limited oscillation swing. High-value inductors result in low FOM at high currents due to the higher kT/C noise and the saturation of the amplitude growth with current increase. The peak figure of merit can be shown to be proportional to Q^2 and is independent of the value of the inductance. Therefore various inductors reach different FOM depending only on

FIGURE 13 *VCO figure of merit in dB using various library inductors.*

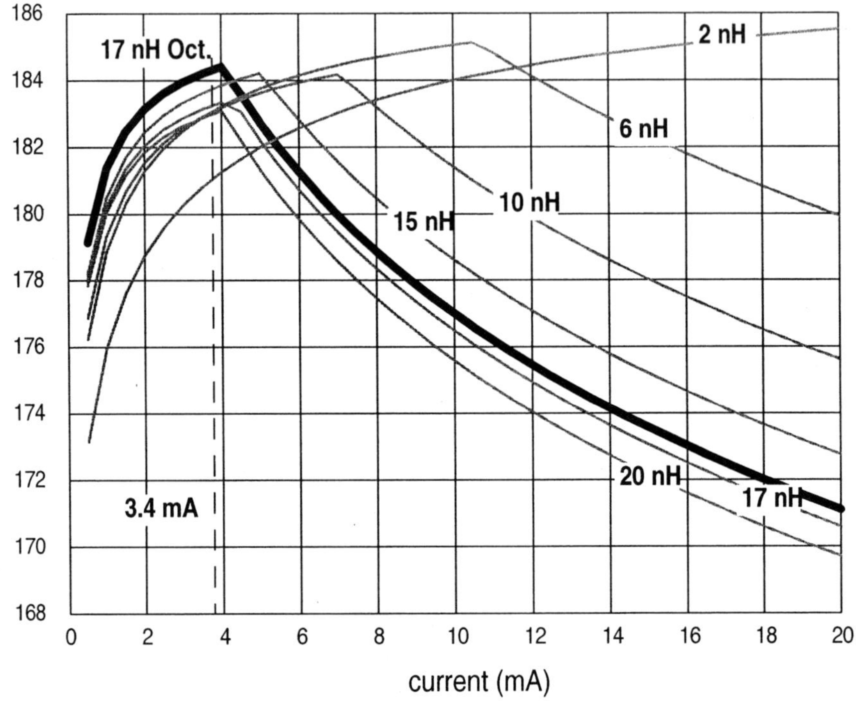

their quality factor and each inductor needs an optimal current to provide peak swing.

Once the output swing is set, the overdrive voltage of the current source device can be optimized. Using (3), the single ended oscillation amplitude is 2.1 V, which means that the current source has headroom of 0.4 V. This is the exact headroom we assume for the noise calculation. That means that our circuit optimization is almost over and the rest of the optimization can be done on the layout level.

The choice of the switching pair device size is almost arbitrary. The use of minimum feature transistors is best because it results in least parasitics. Though a rigorous relationship between the f_T of the switching transistors and their phase noise contribution has not been established to date, simulations show that it is better to go for shorter channel devices. The current source device can be implemented using longer channels as shown later.

This oscillator is fabricated in 0.35 μm CMOS process from STMicroelectronics. The process has 4 metal layers and the inductor is implemented as a parallel stack of metal 4 and metal 3. The current source is implemented as a group of square gate transistors as shown in Figure 14. Phase noise is measured using frequency discrimination method. The resulting phase noise is shown in Figure 14.

FIGURE 14 *Measured phase noise of the VCO.*

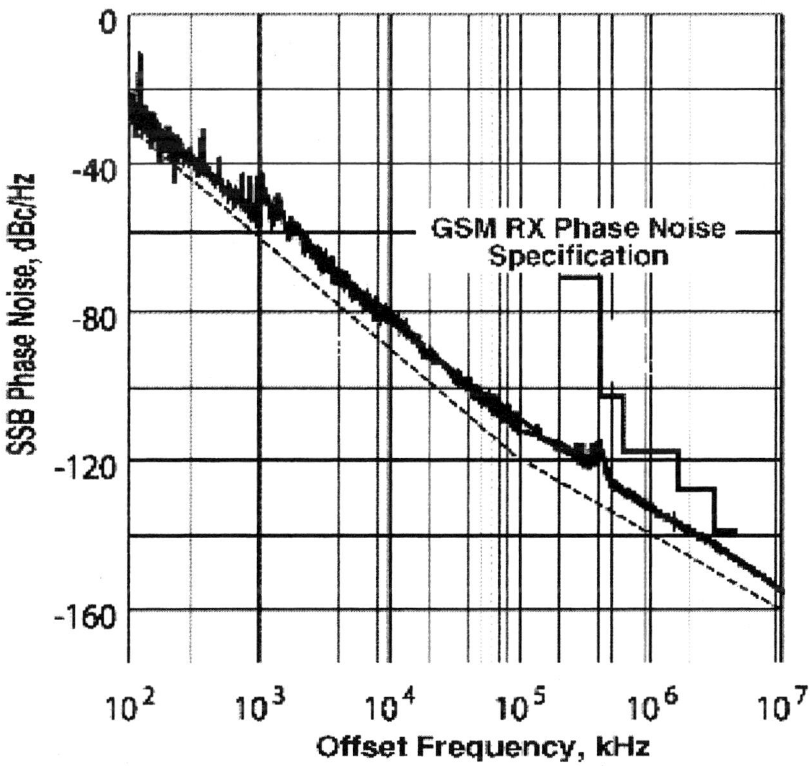

6 Intuitive Explanation of Phase Noise Sources

As already shown, the noise factor, F, for the differential LC CMOS oscillator is given by:

$$F = 1 + \frac{4\gamma RI}{\pi V_o} + \gamma\frac{4}{9}g_{m\ bias}R \tag{12}$$

where I is the bias current, γ is the channel noise coefficient of the FET (equal to 2/3 for a long channel FET, and larger than that for shorter channels) and $g_{m\ bias}$ is the transconductance of the current source FET. The expression in (12) describes three noise contributions: from the tank resistance, the differential pair FETs, and from the current source. Note that thermal noise in the differential pair FETs produce phase noise that is independent of the FET size. In typical oscillators operating at high current levels with moderate to high resonator quality factors, the current source contribution dominates other sources of phase noise. Suppose that the current source contribution, the third term, is removed from (12). Since (3) states that the oscillation amplitude is directly proportional to the current I, the oscillator noise factor simplifies to its minimum value:

$$F = 1 + \gamma \tag{13}$$

This can be understood by considering the circuit in steady state. A cross-coupled differential pair biased at the balance point by an ideal noiseless current presents a negative resistance g_m, with a noise voltage spectral density of $4kT\gamma g_m$ due to the FET pair. However, if the differential pair is switched to one side or the other, there is no differential output noise. Therefore, the oscillation samples the FET noise at every differential zero crossing, that is, at twice the oscillation frequency. Meanwhile, the fundamental component of the differential pair current relative to the oscillation voltage presents a steady state negative conductance exactly equal to the resonator loss conductance. Figure 15 shows an equivalent circuit representing these two actions. [18] analyzes sampled noise in the switched differential pairs of mixers. Although the noise is cyclostationary [19], its spectrum is white. The analysis shows that the noise spectral density out of the sampled transconductance current is equal to that of a linear resistor, $2R$, but with a noise factor of γ. Referring to

Figure 10b, it is now evident how (13) arises. In a sensibly designed oscillator circuit, the third term in (13) can raise the noise factor typically by 75%.

The proof of the model in Figure 15 is not very difficult. The following analysis is based on the analysis performed by Darabi [18] for CMOS mixers. For simplification, we consider the switching pair to be fast switching. The output current is switched between I and $-I$ within a time of $\Delta V/S$ where S is the slope of the differential oscillation waveform at the zero crossing and ΔV is twice the differential voltage required to fully switch the cross-coupled pair. As shown in Figure 16, the differential transconductance of the pair can therefore be approximated in time by a sequence of square pulses of height $-2I/\Delta V$ and width of $\Delta V/S$. The time average of this transconductance is given by

$$G_{m_{avg}} = -\frac{2I}{ST} \tag{14}$$

where T is the period of oscillation and the slope S is given by

$$S = A\omega_o \tag{15}$$

where A is the single-ended oscillation amplitude.

$$G_{m_{avg}} = -\frac{2I}{A\omega_o T} = -\frac{2I}{\frac{2IR}{\pi}2\pi} = -\frac{1}{2R} \tag{16}$$

Therefore, the average transconductance of the switching pair is equal and opposite in sign to that of the total tank loss, $2R$. This is exactly what one would expect for a constant amplitude oscillation where there is no excess negative resistance beyond the tank loss.

Noise associated with the switching pair is treated in the same manner. The transistors produce an input referred noise spectral density of

$$\overline{v_n^2} = \frac{4kT\gamma}{G_m} \tag{17}$$

Again, with the aid of (3), (14), and (15) it is easy to prove that the effective output noise current is equal to

$$\overline{i_n^2} = \frac{4kT\gamma}{2R}, \tag{18}$$

FIGURE 15 *Model of the switch's noise.*

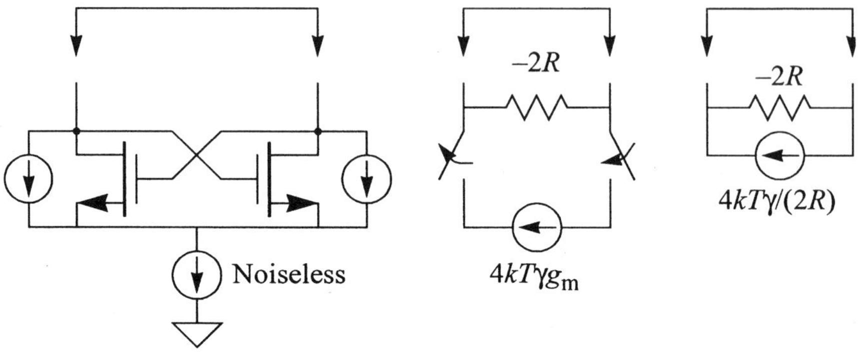

which is equivalent to that of a $2R$ resistor magnified by γ.

The only remaining part of this analysis is to prove that the sampled white noise of the transistors is white and cyclostationary. Since the transistor noise is white and the switching action of the transistors does not entail energy storage, the sampled noise at any time instant is uncorrelated with the noise at any other instant in time and therefore the sampled noise is still white. The power spectral density of the sampled noise is indistinguishable from that of the unsampled noise [18]. Note that the preceding analysis is correct only when the amplitude of oscillation is much larger than ΔV, which is always the case in differential oscillators and mixers.

7 Loading in Current-Biased Oscillators

First consider the differential LC oscillator where the current source is replaced by a low impedance to ground, in the extreme case a short circuit (Figure 17a). Let us take a closer look at the role of the two transistors across the oscillation cycle. First note that the oscillator topology forces V_{GD} of the two FETs to be equal in magnitude but with opposite signs to the differential voltage across the resonator. At zero differential voltage, both switching FETs are in saturation, and the cross-coupled transconductance offers a small-signal negative differential conductance that induces startup of the oscillation. As the rising differential oscillation voltage crosses V_T, the V_{GD} of one FET exceeds V_T, forcing it into the triode region and the V_{GD} of the other FET falls

FIGURE 16 *Transconductance of the switching pair.*

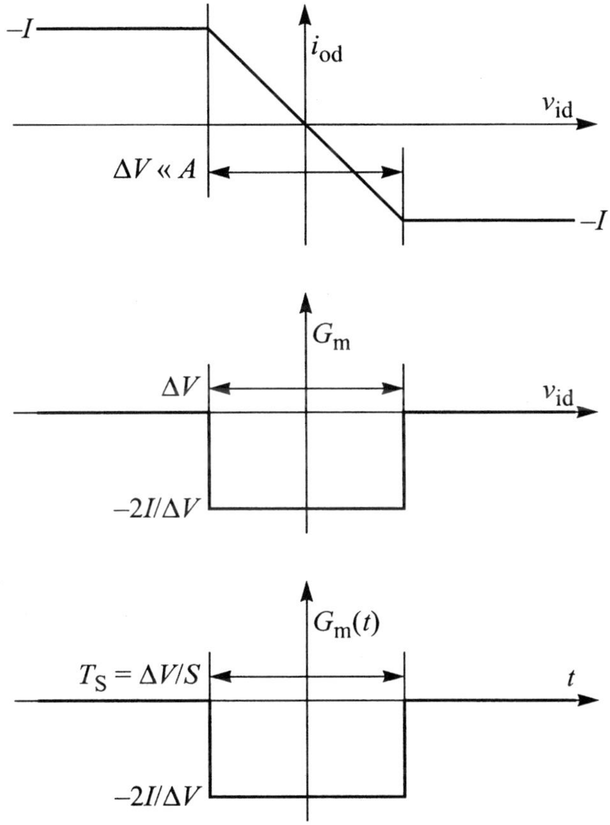

below V_T, driving it deeper into saturation. The g_{DS} of the FET in triode grows with the differential voltage, and adds greater loss to the resonator because the current flowing through it is in-phase with the differential voltage. In the next half cycle, g_{DS} of the other FET adds to resonator loss. The two FETs lower the average resonator quality factor over a full oscillation cycle.

Now suppose an ideal noiseless current source is present in the tail of the differential pair, as shown in Figure 17b. Close to zero differential voltage, the two FETs conduct and present a negative conductance across the resonator. Suppose that in the balanced condition when each FET carries $I/2$, the W/L

ratio is chosen such that $V_{GS} - V_T < V_T$. When the differential voltage drives one FET into triode, it turns off the other FET. As no signal current can flow through the g_{DS} of the triode FET, this FET does not load the resonator, thus preserving quality factor of the unloaded resonator. The differential pair injects noise into the resonator only over the short window of time while both FETs conduct. Referring to the earlier analysis, this means that the noise factor is $1+\gamma$, which is the fundamental minimum for a practical oscillator.

FIGURE 17 *Role of the current source: (a) no current source, and (b) ideal noiseless current source.*

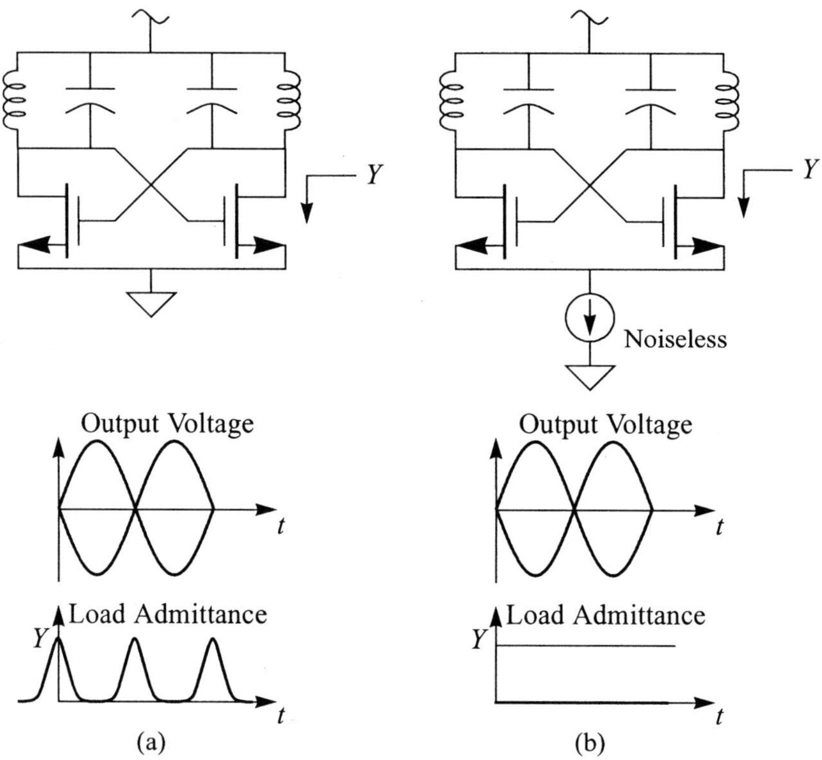

8 Sizing of the Current Source Device

As it appears from (12), the contribution of the switches to phase noise is independent of transistor sizing, to a first order at least. The contribution of the current source, however, is dependent on the sizing and biasing of the current source transistor. It is clear that the smaller the product of the current source transconductance and the tank resistance, the smaller the contribution of the current source to the oscillator noise factor, F. Reducing the effective tank resistance, R, is not an option once the power budget and supply are set. The reduction of the current source device transconductance is a more viable alternative. This transistor is not required to have any "dynamic" behavior and therefore, it need not have high transconductance. Increasing its V_{GS} is therefore an effective way of reducing its transconductance. However, the transistor needs to remain in saturation throughout the cycle to operate as a current source. A higher V_{GS} pushes the current source transistor into triode at a lower current level and hence limits the maximum amplitude.

To gain more insight, let's plot the oscillator phase noise versus current in light of (12). We will assume the tank quality to be constant as well as the tank inductor and capacitor. In other words, for an already designed oscillator we will sweep the current and look at the resulting phase noise. The result is shown in Figure 18. In the current-biased regime, phase noise drops with the increase in bias current at a rate of approximately 20 dBc/Hz/A. The breakpoint in the curve occurs when the current source transistor enters triode allowing the switches to load the tank and reduce its average quality factor. This effect balances the amplitude increase and makes phase noise constant despite the increase in current. For a current source device with a larger W/L, the overdrive is smaller and therefore, the lowest "available" phase noise is lower than that with a larger W/L. In addition, larger size for the current source transistor reduces the device flicker noise, which is inversely proportional to the transistor area.

The size of the current source transistor cannot be arbitrarily large. The reason for that is the parasitic capacitance that is inevitably incurred with a larger size. The large parasitic capacitance gives a lower impedance path of the even harmonics to ground. This, again, allows the triode switches to load the tank and reduce the apparent average quality factor, degrading phase noise performance, much like the voltage-biased oscillator. The other problem with a very

FIGURE 18 *Phase noise versus current.*

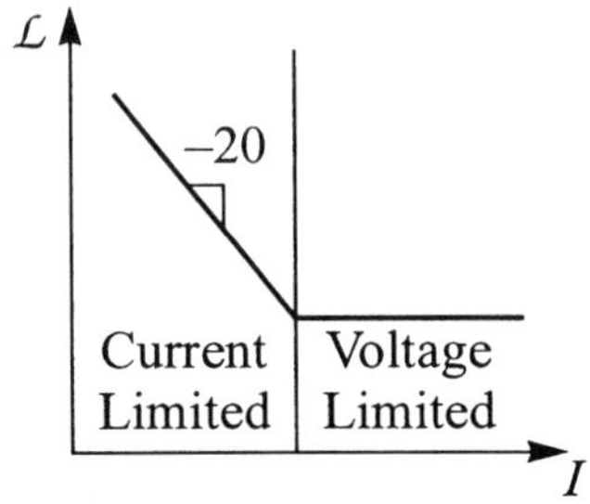

large current source transistor size is flicker noise up-conversion. It is true that the larger size reduces the amount of flicker noise however, flicker noise appears around the oscillation frequency *only* through up-conversion typically through some FM modulation process. Higher parasitic capacitance at the common mode point increases flicker noise up-conversion gain from the switching transistors. the only solution known to this paradox is through the use of noise filtering technique to break the trade-off [20]. A more comprehensive discussion of the noise filtering technique is given later in this chapter. At this point it is enough to point out that the capacitance at the common mode point can be reduced by better layout. In particular, through the use of ring-shaped transistors. A ring shaped device, as shown in Figure 19, has an effective width equal to its centerline perimeter [21] and has the smallest drain area achievable while the source area is highest. For a grounded source transistor used in the current source, the use of ring-shaped transistors is the best choice. If rings are not allowed by design rules, a square approximation is still very helpful.

9 Noise Filtering in Oscillators

9.1 Role of the Current Source

The mechanistic analysis in [22] shows how current source noise creates phase noise in the oscillator. The switching differential pair, which acts as a single-balanced mixer for noise in the current source, commutates and up-converts low frequency noise (modeled by a single tone) into two correlated AM sidebands around the fundamental. Therefore, low frequency noise in the

FIGURE 19 *Layout of the current source device.*

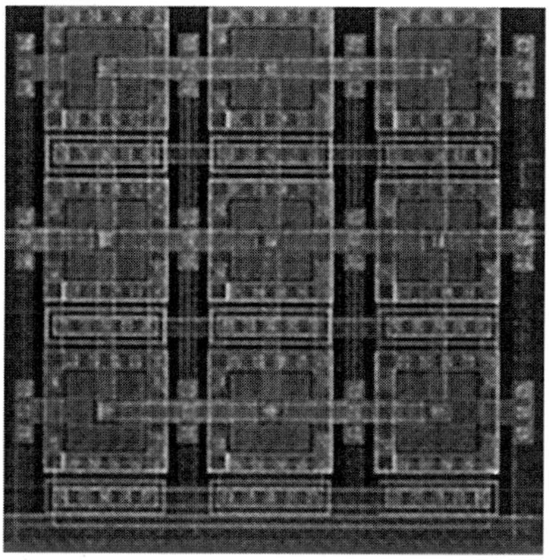

current source does not directly produce phase noise. Noise frequencies around the fundamental tone are translated far away from the passband of the tuned circuit. However, noise frequencies around the 2nd harmonic down-convert close to the oscillation frequency, and up-convert to around the 3rd harmonic where they are rejected by bandpass LC tank. A tone injected into the tuned circuit passband may be decomposed into half AM and half PM sidebands around the oscillation frequency [23]. Thus, half the noise in the current source lying at frequencies close to the 2nd harmonic produces phase noise. Odd harmonics of the commutating waveform might down-convert higher noise frequencies close to the oscillation, but for a first-order calculation these are ignored because the mean square contributions fall off rapidly ($1/3^2$, $1/5^2$, $1/7^2$,...). The role of noise around even harmonics of the oscillation was first noted based on the similarity between the switching pair of an oscillator and a commutating mixer [24] and later based on an analysis of the time-domain waveforms [25].

The current source plays a two-fold role in the differential LC oscillator: it sets the bias current, and it also inserts a high impedance in series with the switching FETs of the differential pair. In any balanced circuit, odd harmonics circulate in a differential path, while even harmonics flow in a common-mode path, through the resonator capacitance and the switching FETs to ground. Therefore, strictly speaking the current source need only provide high impedance to even harmonics of the oscillation frequency, of which the 2nd

FIGURE 20 *Current source impedance requirements.*

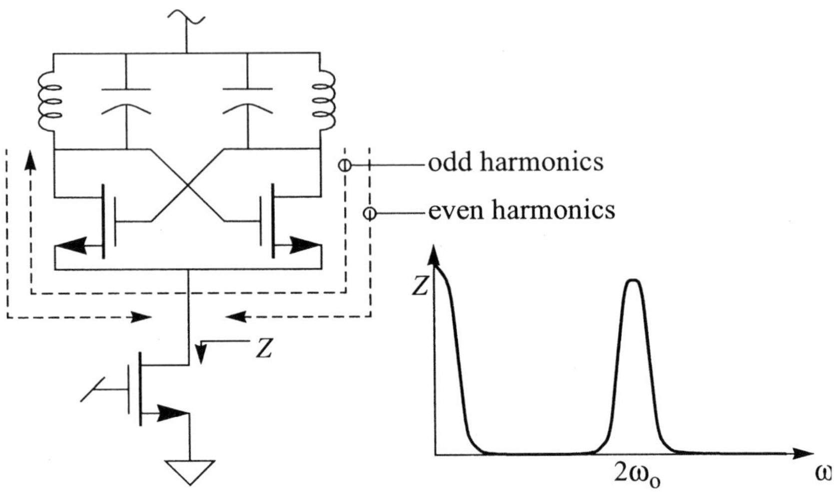

harmonic is usually dominant. Shrinking the requirement of high impedance to a narrow band of frequencies offers some unique opportunities to realize this concept.

9.2 Noise Filtering

To recapitulate, for the current source:

1. Only thermal noise in the current source transistor around the 2nd harmonic of the oscillation causes phase noise, and
2. a high impedance at the tail is only required at the 2nd harmonic to stop the differential pair FETs in triode from loading the resonator.

This suggests use of a narrowband circuit to suppress the troublesome noise frequencies in the current source-making it appear noiseless to the oscillator. This gives a high impedance in the narrow band of frequencies where it is needed. Placing a large capacitor in parallel with the current source, as shown in Figure 21a shorts noise frequencies around $2\omega_0$ to ground. Then, to raise the impedance, an inductor is inserted between the current source and the tail, as shown in Figure 21b.

FIGURE 21 *Tail-biased VCO with noise filtering: (a) capacitor alone, and (b) complete topology.*

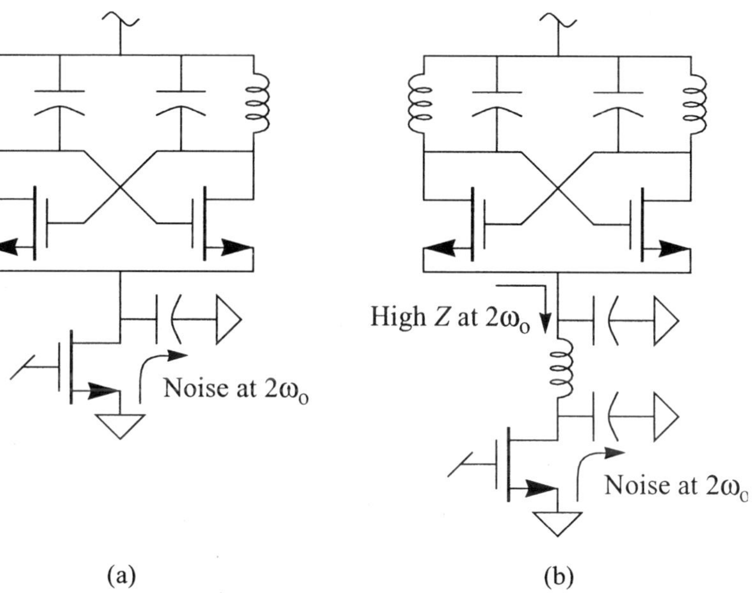

The inductance is chosen to resonate at $2\omega_0$ in parallel with any capacitance present at the common sources of the differential pair. The impedance at the tail is limited only by the quality factor of the inductor. The inserted inductor and the large capacitor comprise what we term a noise filter.

A variant of the tail-biased LC oscillator described in literature is the top-biased differential LC oscillator, where the current source is connected from the positive supply to the center tap of the differential inductor (Figure 22). If the junction capacitors to ground are ignored, these two oscillators are identical in that the bias current source is in series with the supply voltage source, and the position of the two can be exchanged without affecting the circuit topology. However, the two circuits are different when the junction capacitors are included. These differences have some practical consequences. For instance, the top-biased oscillator is more immune to substrate noise because the current source is placed in an *n*-well, rather than in the substrate. Also, the top-biased oscillator up converts less flicker noise into phase noise.

FIGURE 22 *Top-biased VCO with noise filter.*

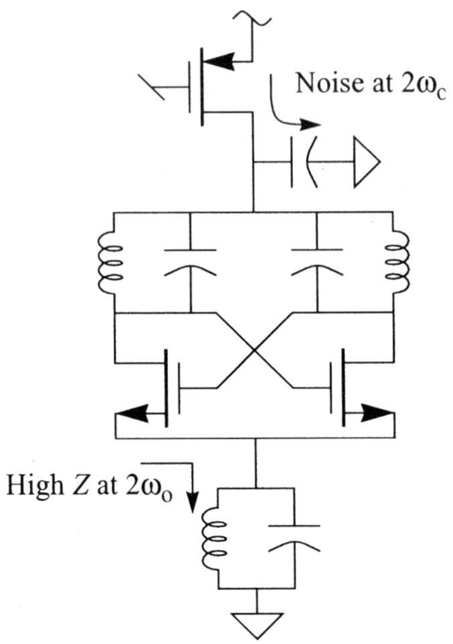

One important consequence of the difference is in the noise filter circuit for the top-biased oscillator. As before, a large capacitor in parallel with the current source shunts noise frequencies around the 2nd harmonic to ground. However, the filter inductor must be inserted at the common source point of the switching pair to resonate in parallel with the capacitance at that node at $2\omega_o$. This blocks 2nd harmonic current from flowing through the grounded junction capacitors comprising the resonator, and through the switching FETs to ground.

A properly designed noise filter brings the noise factor of the differential LC oscillator down to its fundamental minimum of $1+\gamma$. Of course this is assuming the filter inductor to have high-enough Q. Once the constant of proportionality F is minimized, (12) prescribes that, given a resonator Q and current limited operation, for least phase noise the oscillation amplitude V_o must be as large as possible. The positive peak of the oscillation at the drains of M_1, M_2 is limited by breakdown. The negative peak can, in principle, be as low as a forward-biased junction voltage below ground. The instantaneous negative voltage is absorbed across the filter inductor, and the large filter capacitor holds the V_{DS} constant across the current source FET, maintaining it in saturation. In practice, the highest differential voltage swing across the LC resonator is roughly $2V_{DD}$ peak.

What was ignored in (13) is the noise of the filter inductor. This noise has the same transfer function as noise from the current source transistor before without the noise filter. Therefore, the actual noise factor with the filter present is:

$$F \approx 1 + \gamma + \frac{4}{9}\frac{R}{Q_f 2\omega_o L_f} + \frac{4}{9} g_{m\ \text{bias}} R \left(\left(\frac{2\omega_o}{\omega_f} \right)^2 - 1 \right)^{-2} \tag{19}$$

Clearly, using a low quality inductor in the noise filter defeats the purpose of lowering common mode noise originating around $2\omega_o$ in the current source. No matter how low the quality factor is, the current source noise will be eliminated however, low quality inductors can be even noisier if poorly selected.

In the extreme, if a very large gate bias is applied to the current source FET, it is continuously in triode region and almost appears as a short circuit to ground (an example is given in [26]). This reverts to the voltage-biased oscillator (Figure 23). Without a noise filter, the FETs will load the resonator throughout the oscillation while the differential resonator voltage is larger than V_T.

FIGURE 23 *Voltage-biased VCO with the filter.*

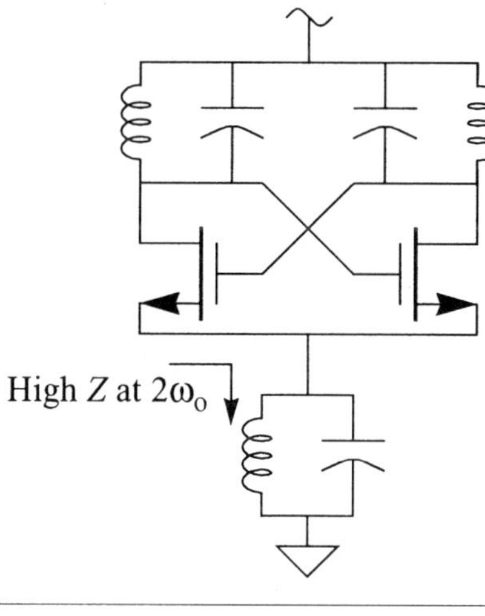

A noise filter in the tail tuned to the 2nd harmonic removes this loading (Figure 23). This circuit oscillates with the largest possible amplitude because there is no current source FET in series with the differential pair to take up voltage headroom. As there is no 2nd harmonic with DC content, the circuit biases at a current that relates to the differential oscillation amplitude according to (6) because whatever this current is, it will be commutated into the resonator by the periodic FET switching. In other words, with the noise filter the circuit acts as if biased by a tail current source that consumes zero voltage headroom, yet produces the largest possible amplitude. As a result the phase noise of this topology is least, although its current is largest.

10 Prototype Oscillator

It was the intent of this work to show how a noise filter can lower phase noise to a record level in a prototype CMOS oscillator with fully on-chip resonator. As such F must be low (the task of the noise filter), the amplitude V_0 should be large, and the inherent quality factor of the resonator, Q, must also be large. The latter two yield quadratic improvement in phase noise, and are also related in that the higher the resonator Q, the higher the amplitude for a given bias current. In an entirely integrated resonator, losses in the on-chip spiral inductor usually limit Q.

Before describing the design of the spiral, it is necessary to first specify the IC fabrication process. The ST Microelectronics BiCMOS6M process used here offers four layers of metal, and a substrate resistivity of 15 Ohm-cm. The Metal 4 film is almost 2.5 μm thick, while the lower layers of metal are all about 0.8 μm thick.

The inductor is designed using a custom fast simulator that models self-inductance, parasitic capacitance, and all losses, including dissipation due to displacement and eddy currents in the substrate [16]. The differential resonator uses a single balanced octagonal spiral with a center tap (Figure 24). Although the obvious geometrical symmetry of this structure is discussed in the literature [27], we believe its main benefit is that it offers higher Q than two independent spirals in series, each of half the required inductance. For the same dimension of the inner hollow area, the differential spiral requires roughly 1/2 the length of metal of two spirals in series, which lowers metal loss, and it occupies a smaller footprint over the substrate, which lowers substrate losses.

FIGURE 24 *Differential inductor.*

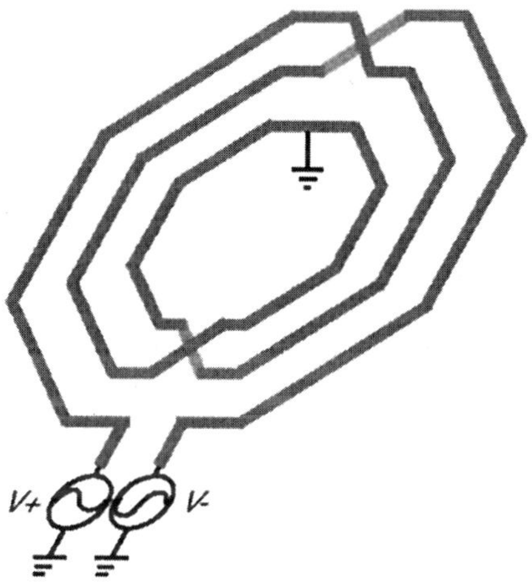

The balanced spiral in Metal 4 implements a total differential inductance of 26 nH, and is optimized for 1.1 GHz. At this frequency, simulation shows that displacement current loss is important in the lightly doped substrate. Simulations also show that a Metal 1 patterned shield is more effective in blocking displacement currents from entering the substrate than a polysilicon shield [17]. However, the shield geometry must be designed very carefully because the high conductivity of metal, which improves shielding of the electric field, makes it more susceptible to strong eddy currents induced by the magnetic field under the inductor. The shield is patterned as a checkerboard of small squares connected diagonally with narrow tracks, which locally cancel the magnetic fields induced by miniature loops of current within the squares (Figure 25). The patterning ensures there is no closed loop of metal at the scale of the inductor, which might allow image current to flow. The simulated Q of this inductor at 1.1 GHz is about 13, which is verified by measurement.

A tail-biased differential oscillator is implemented in 0.35μm CMOS. The oscillator consumes 3.7 mA from 2.5 V supply. Measured phase noise 3 MHz away from a 1.2 GHz oscillation is –153 dBc/Hz as shown in Figure 26. A

FIGURE 25 *Shield patterning.*

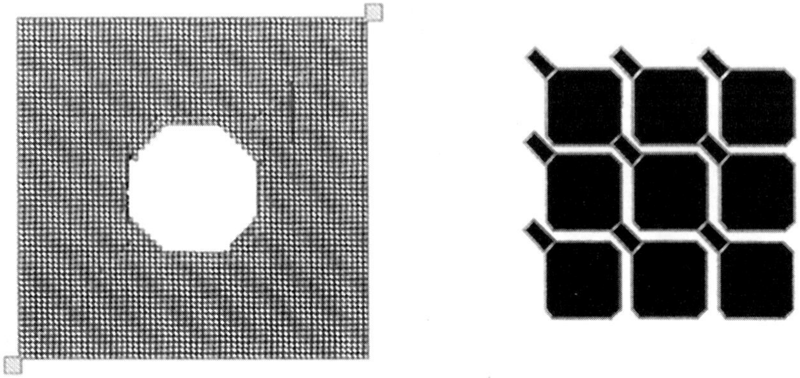

reference oscillator, which is identical except it has no noise filter, is fabricated on the same wafer. Its phase noise at 1.2 GHz and the same offset is 7 dB worse.

A top-biased oscillator was also fabricated on the same wafer. This circuit tunes from 1 to 1.2 GHz and also consumes 3.7 mA. Measured phase noise at 3 MHz offset is –152 dBc/Hz, an 8 dB improvement over its reference oscillator, (Figure 27). The noise filter uses a 10 nH square on-chip spiral and a 40 pF MiM capacitor. A third oscillator implemented in the same process oscillates at 2.1 GHz. Consuming 4 mA from a 2.7 V supply, its measured phase noise at 3 MHz offset is –134 dBc/Hz. The 5.5 nH on-chip differential inductor has a Q at 2 GHz of about 10.

The phase noise of a commercially available discrete transistor module oscillator, the Vari-L VCO190-1100AT, that tunes over the same frequency range was measured on the same instrument, and is plotted alongside. The VCO core of the module is estimated to consume 1/3rd the total current, which is roughly the same current as the 1 GHz VCO's described above, but from a 5 V supply. Measured phase noise at 1 GHz and 3 MHz offset is –150 dBc/Hz (Figure 26).

Now let us use Leeson's proportionality combined with our definition of the noise factor to manually calculate this oscillator's phase noise. The foundry specifies $\gamma = 4/3$ for the FETs, and the various other parameters are

FIGURE 26 *Tail-biased VCO phase noise measurement.*

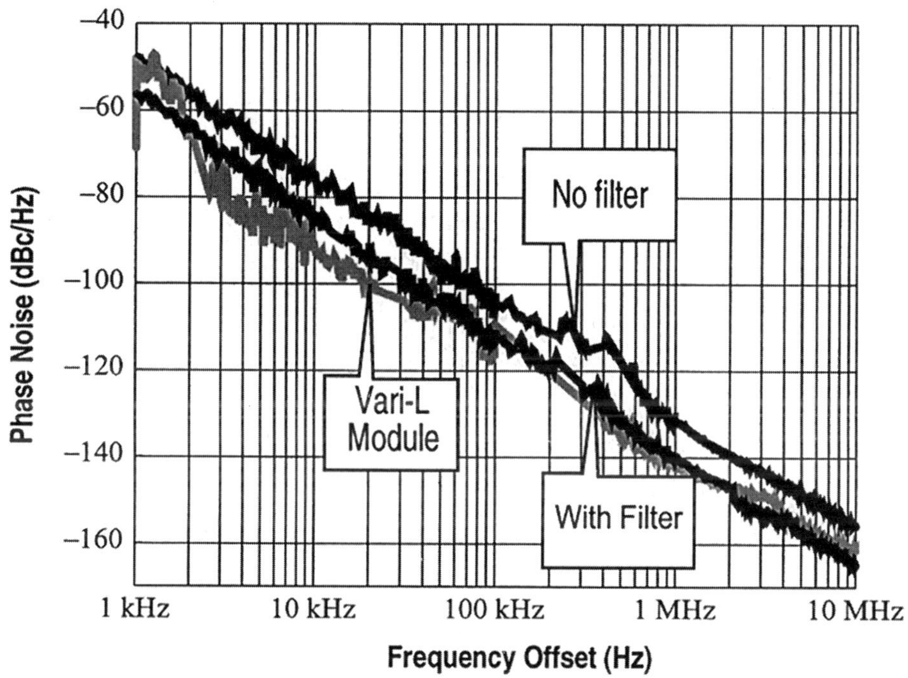

$L = 13$ nH, $Q = 14$ at 1.2 GHz, $I = 3.7$ mA, $f_o = 1.2$ GHz, $f_m = 3$ MHz, and $V_{DD} = 2.5$ V. The calculated $\mathcal{L}(f_m) = -153.2$ dBc/Hz is remarkably close to the measured value. This further validates the mechanistic phase noise analysis presented earlier.

11 Practical Considerations

11.1 Power Supply Rejection with Noise Filtering

Does the noise filter adversely affect power supply rejection in this differential oscillator? Consider the tail-biased differential oscillator (Figure 2) without noise filter or parasitic capacitances. If the varactor is biased with respect to V_{DD}, fluctuations at any frequency in V_{DD} with respect to ground cannot modulate the varactor, creating no FM. Furthermore, the current source in the

FIGURE 27 *Phase noise measurement for the top-biased VCO.*

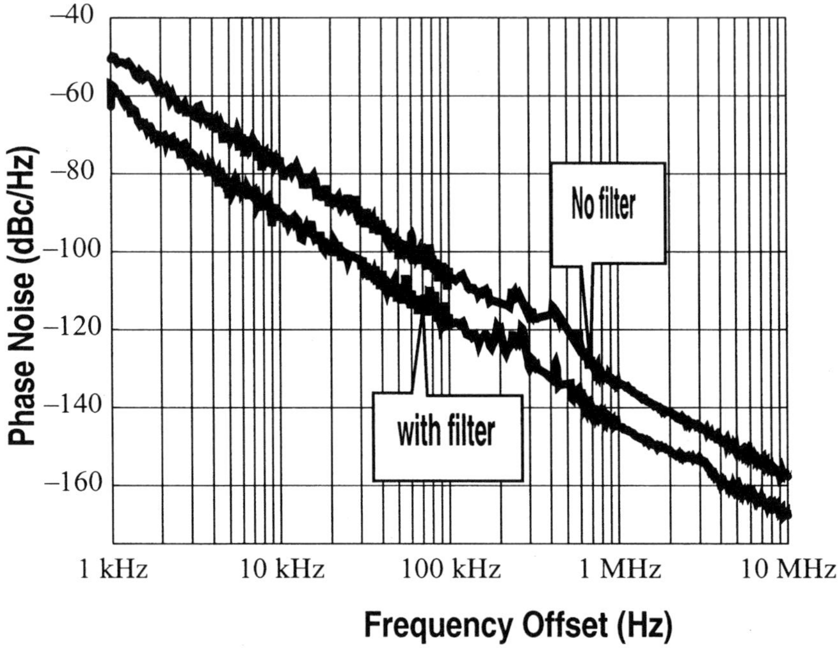

tail when biased at constant V_{GS} will not permit any AM. However, low frequency noise on V_{DD} passes through the resonator inductors to modulate the voltage on the FET drains, and voltage-dependent junction capacitors to ground there will create FM noise. An analysis of this effect was given in the previous chapter.

With the filter present (Figure 21b) the oscillator responds in exactly the same way to low frequency noise on the supply. At high frequencies the filter capacitor bypasses the current source FET and couples supply fluctuations directly into the filter inductor.

The oscillator can only respond with a common-mode current, which as explained earlier must be either around zero frequency or at even harmonics. But the filter inductor in parallel resonance with the tail capacitance blocks any 2nd harmonic. In this way the oscillator better rejects noise on the power supply. The same analysis applies to the top-biased oscillator, where the

FIGURE 28 *Switching FET noise conversion gain at 10 kHz.*

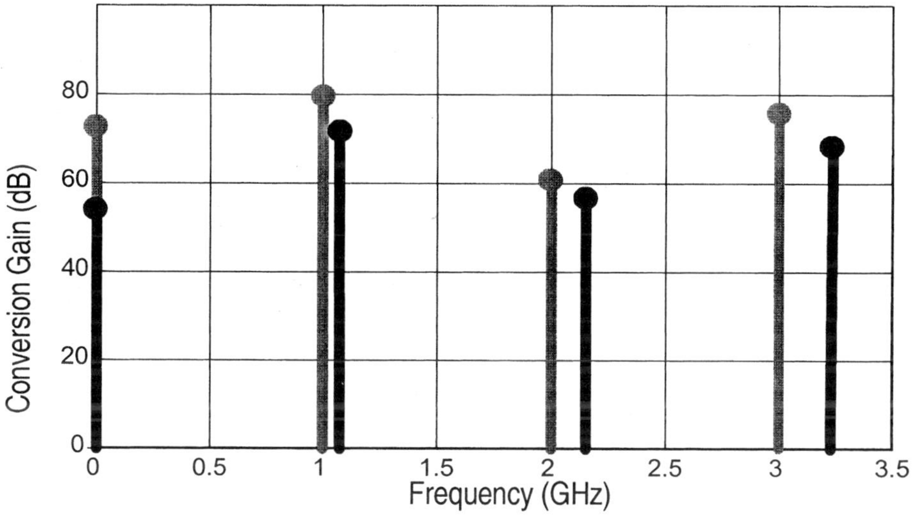

grounded filter capacitor shunts power supply noise and the series inductor blocks noise currents.

Cadence's SpectreRF is found to be a useful and accurate aid in understanding mechanisms of phase noise. It allows the simulation of conversion gain from voltage noise at any node in the oscillator to phase noise. Figure 28 plots the simulated conversion gain from gate-referred voltage noise in one of the differential pair FETs to phase noise, versus discrete frequencies of voltage noise that will produce close-in phase noise. The noise filter reduces up-conversion of low frequency noise, but does not affect the noise at harmonics of the oscillation frequency as much.

As predicted, this simulation shows that noise in the differential pair FETs close to the oscillation frequency is mainly responsible for phase noise. Figure 29 shows that the noise filter dramatically lowers the conversion gain of noise in the current source at frequencies around the 2nd harmonic. Finally, Figure 30 shows that the noise filter improves power supply rejection at all frequencies.

In sum, these simulations verify the qualitative description in Section 9.2 that with the filter present, only noise in the differential pair and the resonator loss

FIGURE 29 *Current source FET noise conversion gain at 10 kHz.*

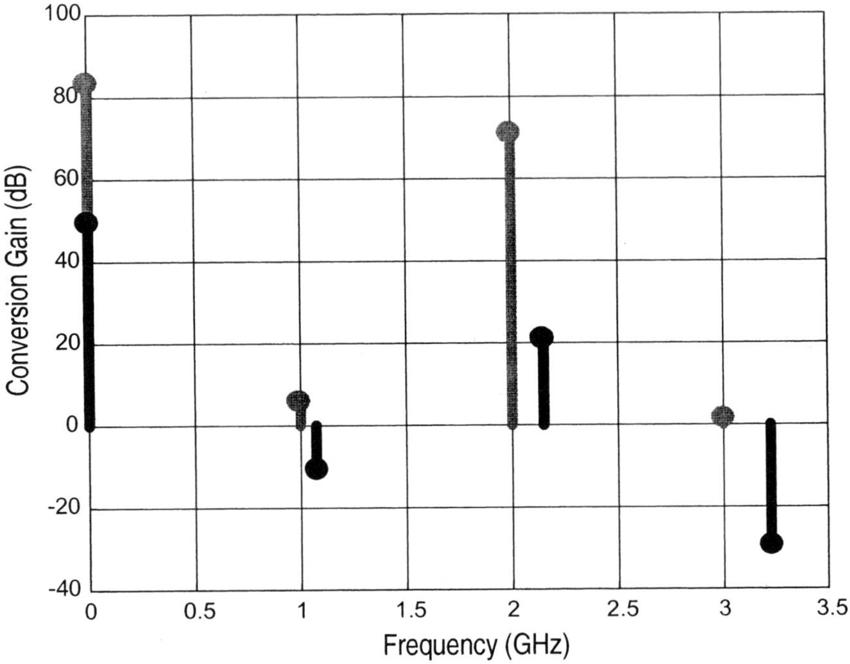

matters. As a side note, when 2nd harmonics no longer flow in the resonator, the oscillation frequency increases slightly [6].

11.2 Device Limitations on Maximum Swing

In a tail-biased oscillator, the single ended oscillation is centered on a common mode level equal to the supply voltage. Being a large signal, the drain of the switching transistors inevitably rises to large voltage levels that are not appropriate for reliable continuous time operation [28]. As a rule of thumb for most technologies, the maximum voltage on the transistor cannot exceed 1.5 V_{DD} where V_{DD} is the maximum allowed supply voltage. This limits the maximum oscillation amplitude to 0.5 V_{DD}, which is half the maximum achievable or 6 dB higher phase noise than the achievable minimum. Some processes are even more stringent and the limit on maximum swing is set at 1.25 V_{DD}.

FIGURE 30 *Power supply noise conversion gain at 10 kHz.*

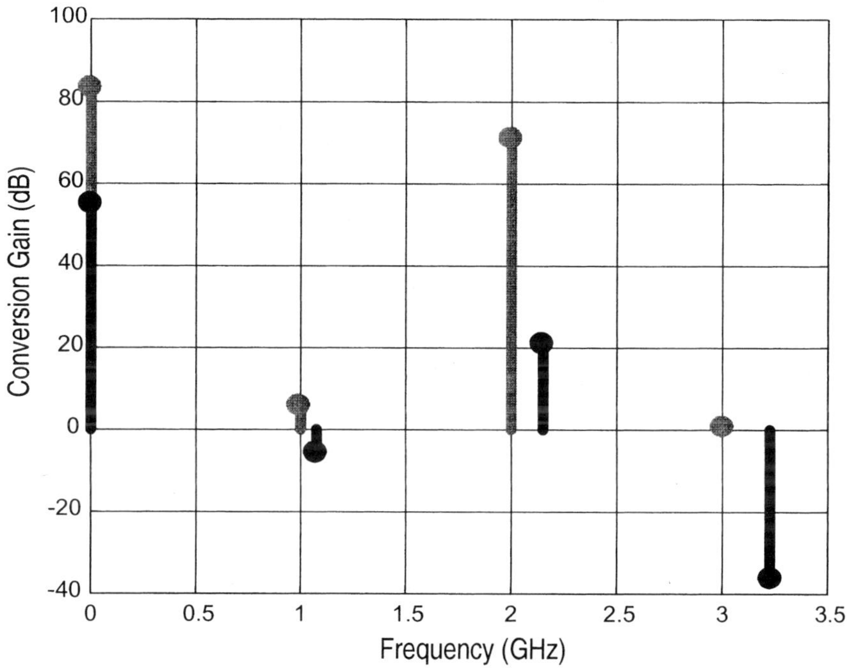

From a reliability perspective, the complementary differential topology is better than a typical tail-biased oscillator because the common mode level of the oscillation is situated at a voltage level below the supply giving more headroom for swing. If not for any other reason, this fact gives the complementary differential topology a performance advantage over tail-biased topology. With a noise filter in place this advantage is even more eminent. The best arrangement for a noise filter for this topology is shown later in Figure 32.

11.3 Extending the Tuning Range

A CMOS oscillator must be designed with a large tuning range to overcome process variations, which create a large spread in the center frequency from wafer to wafer. The simplest way to do so is with a strong varactor, that is, one that gives a large capacitance swing relative to the fixed resonator capacitance as the tuning voltage is swept over its full range limited by the power supply, V_{DD}. Irrespective of how the varactor is realized, whether as a MOS capacitor or a PN junction [21], the larger its area the stronger it is. The resulting high sensitivity of frequency tuning, though, is at the price of worse phase noise, as discussed in the previous chapter.

Varactor capacitance depends continuously on control voltage. Clearly, additive noise on the control voltage will convert through frequency modulation into phase noise sidebands. However, the varactor whose incremental capacitance is a function of the instantaneous voltage across it offers an average capacitance to the resonator that depends on the envelope and duty cycle of the oscillating waveform. As described in Section 9, even if the control voltage is noiseless, the varactor will detect envelope fluctuations due to AM noise on the oscillation and by modulating the average capacitance convert this into FM noise [22][29]. This process may add several dB to the phase noise sidebands.

By combining discrete and continuous tuning, it is possible to lower FM sensitivity while spanning a wide tuning range [21]. For instance, a three-bit binary-weighted switched capacitor array, Figure 31, tunes the oscillator center frequency to 8 discrete frequencies. Then, a small MOS varactor interpolates continuously around these frequencies, giving rise to a family of overlapping tuning curves to guarantee continuous frequency coverage over the tuning range, as shown. This requires a mixed analog-digital PLL, whose design has been described elsewhere [30], to tune such a VCO.

FIGURE 31 *Mixed tuning using switched capacitors.*

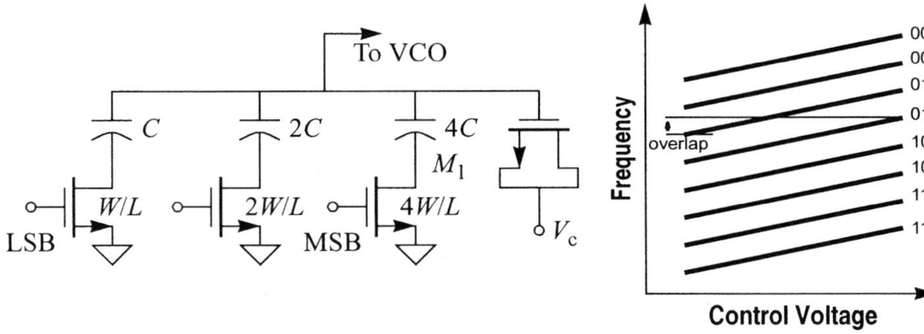

When designing the switched capacitor array, there are two design goals that trade-off against one another: The quality of the switched capacitor arrangement and its tunability. In the on-state, the switch represents a hopefully-small series resistance. Therefore, the quality factor of the switch and the capacitor is given by $Q = 1/(\omega_0 C R_{on})$. The quality factor of the switched capacitor increases with the larger device size due to the reduction in the on-resistance of the switch. The tunability of switched capacitor defined here as the ratio of the on-capacitance to the off-capacitance. When the switch is on, the capacitance of the arrangement is basically that of the switched capacitor. When the switch is off, the capacitance drops to the series combination of the capacitor and the parasitic capacitance of the switch. Therefore, the tunability decreases with the larger device size. Good layout improves the varactor tunability and allows for higher quality switched capacitor. In particular, having a square-gate layout reduces the parasitic drain junction capacitance and in turn, the off-capacitance by a factor of 2~3, which translates linearly to quality factor. In practical oscillators around 1 GHz, the quality factor of the switched capacitor is not limiting the overall tank quality as a Q of 40 is reasonably achievable. At 5 GHz, the achievable Q is one fifth of that (around 8) and can potentially limit the tank quality.

11.4 Noise Filtering in Earlier Works

For completeness, we now summarize previous work suggestive of filtering in oscillators, and compare it with the ideas presented here.

Tchamov and Jarske proposed adding an inductor in series with the current source biasing each delay stage in a ring oscillator, with the object of improving sinusoidal purity of the signal and the output impedance of the current source at high frequency [31]. However, there is no recognition of the inductor's benefits on internally generated phase noise.

Hajimiri and Lee recommended a large capacitor in parallel with the current source, because, they argued, this shrinks the duty cycle of switching current in the differential pair, which lowers the instantaneous FET current at differential zero crossing, thus lowering phase noise due to the differential pair FETs [13]. This idea was implemented by De Muer et al. to lower the up-conversion of flicker noise into phase noise [12]. However, this argument overlooks the effect described in Section 9.2, that the large capacitance at the tail offers a low impedance path for the triode-region FETs in the differential pair to load the resonator and degrade Q. Moreover, the tail capacitance leads to greater up-conversion of flicker noise from the differential pair into oscillator phase noise [22][14]. Our analysis shows that a large filter capacitor alone can only lower thermally generated phase noise in one case. That being when the current source contributes such a large phase noise (3rd term in (12)) that its elimination by the capacitor outweighs the higher phase noise due to degraded Q. We believe that in practice this is unlikely.

11.5 Noise Filtering in Other Oscillators

It is straightforward to extend the noise filter to other oscillators. The voltage-biased oscillator is improved as shown in Figure 23. Other differential oscillators such as the complementary differential [10] and quadrature oscillators [26] fall into the category of top-biased, tail-biased, or voltage biased, and the appropriate filter reduces their phase noise down to the fundamental minimum.

Figure 32 shows the noise filtered version of the complementary differential oscillator. This topology is the best-suited for commercial implementation. The oscillation is superimposed on a bias of $V_{DD}/2$ resulting in best reliability. Moreover, the current switching efficiency is better than all other topologies including Colpitts oscillator that never reaches its theoretical efficiency. Adding the noise filter as shown lowers the noise figure of this oscillator to the minimum. This noise-filtered topology can indeed give 3 dB improvement over the regular tail-biased oscillator with a noise filter. The noise factor of this topology is given here without proof as

$$F = 1 + \gamma_n + \gamma_p . \qquad (20)$$

Figure 33 shows a noise filter at the bias current source in a Colpitts oscillator. The capacitor C_2 should remain unaffected by insertion of the filter, which means that additional capacitance may be required to cancel the inductor reactance at the oscillation frequency. Here, given the single-ended FET circuit, the filter inductor tunes the parasitic capacitance to the oscillation frequency instead of the 2nd harmonic as in differential oscillators. SpectreRF simulations confirm that the noise filter lowers phase noise in the Colpitts oscillator.

The commutated-current Colpitts can also make use of the noise filtering technique. Note that one objective is to deliver a high impedance at f_o to the Colpitts transistors M_1 and M_2 (Figure 34). The other objective is filter noise

FIGURE 32 *Noise-filtered complementary differential oscillator.*

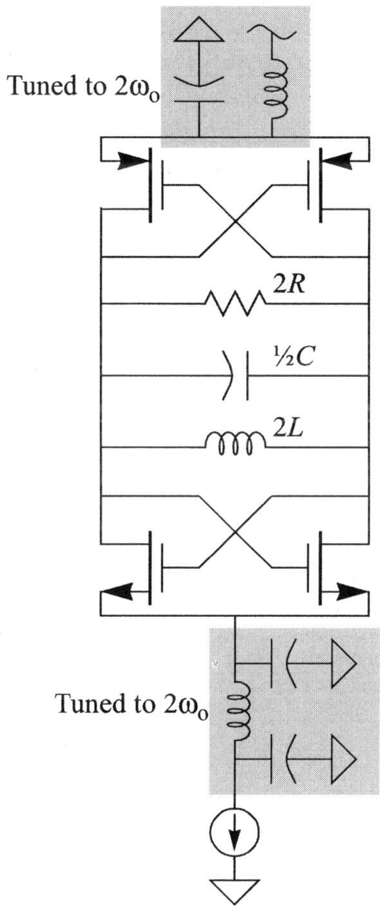

around $2f_o$ from the current source. Inserting an inductor L_f and a large capacitor C_f and tuning the filter inductor with parasitics to provide high impedance at $2f_o$ serves both purposes. The impedance of the noise filter is *sampled* by the commutating switches M_3, and M_4 at the oscillation frequency therefore the filter impedance referred to A or A' will peak at $f_{tune} - f_o$ where f_{tune} is the filter tuning frequency or $2f_o$. Therefore a tank circuit tuned at $2f_o$ at the common mode point provides a high impedance at f_o seen from A or A'. We refer to this filter as the sampled noise filter.

FIGURE 33 *Colpitts oscillator with noise filter.*

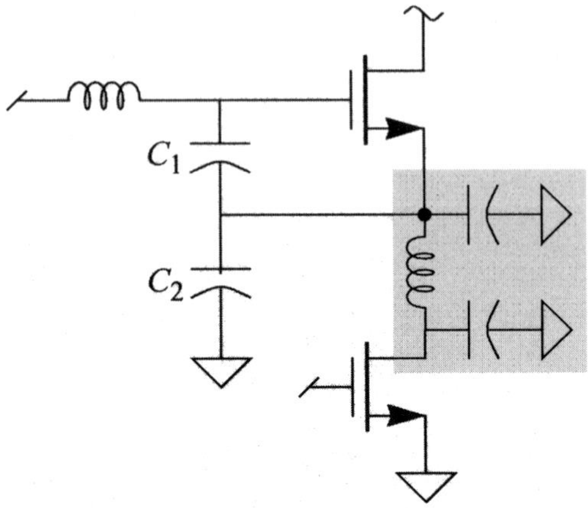

FIGURE 34 *Commutated-current Colpitts and sampled noise filter.*

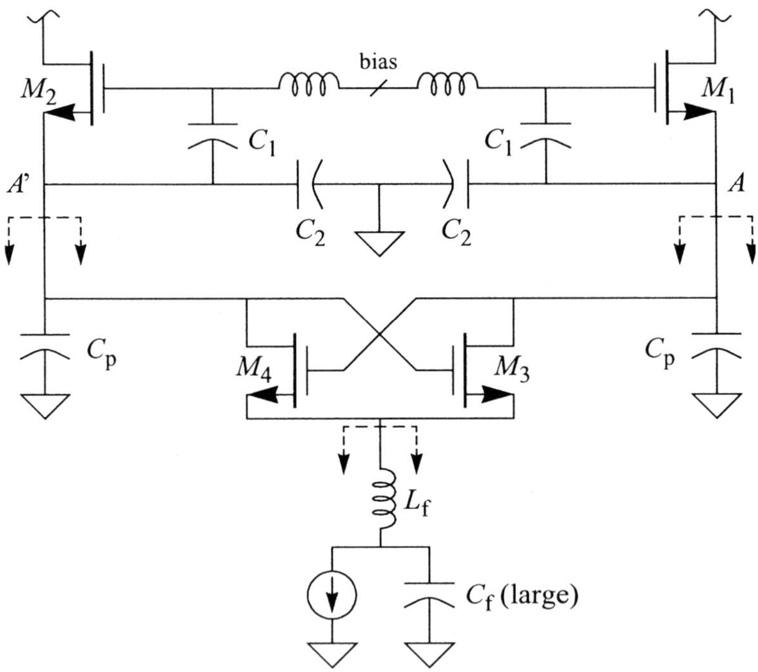

12 Example: Redesign of GSM VCO

In this section, we re-design the GSM VCO presented earlier taking into account all the practical constraints discussed in Section 5. The design implements a noise-filtered tail-biased oscillator (Figure 21b). This time we consider all practical aspects discussed in the previous sections.

First the swing has to be limited to accommodate reliability concerns. This means that the design has a sub-optimal FOM. However, this is necessary for product reliability. We will employ a noise filter so we can assume a noiseless current-source. Now assume the maximum single-ended amplitude allowed for oscillation is $0.65V_{DD}$ (or $1.3V_{DD}$ differentially). Again, just like the design in Section 5, the supply is 2.5 V. We shall use the inductor library shown in Figure 12. Assume that a phase noise of –145 dBc/Hz @ 3 MHz is required to cover process variations. Using (11), and assuming γ of 4/3 we can write the following,

$$QC = \frac{1}{V_o^2} kT \frac{\omega_o}{\mathcal{L}(f)} \frac{2}{\omega^2} F \geq 1.125 \times 10^{-11} . \tag{21}$$

Assuming a tank $Q > 7$ results in a capacitance of around 1.6 pF. This capacitance should tune with the tank inductor at 1.1 GHz. The resulting inductor is about 13 nH. However, the inductor library in Figure 12 doesn't have a 13 nH inductor. The closest values are 10 and 15 nH. The 15 nH inductor will not tune with the specified capacitance at 1.1 GHz, therefore, we use a 10 nH instead. The capacitance that tunes with this inductor is given as:

$$C_{tank} = \frac{1}{\omega_o^2 10 \times 10^{-9}} = 2 \text{ pF} \tag{22}$$

The selected inductor has a Q of 8 and with the capacitor given by (22) we are sure that (21) is satisfied.

The bias current is determined as follows:

$$I_o = \frac{V_o}{\frac{4}{\pi} Q \omega_o L} = 4.6 \text{ mA} \tag{23}$$

This is certainly more than the 3.4 mA we used in the implementation in Section 5 however, this is necessary to ensure the reliability of the design as well as having a comfortable margin over the specifications.

We note that the capacitance given by (22) is the total tank capacitance. This capacitance is divided into three categories, inductor's self-capacitance, varactor tunable capacitance, and parasitic capacitances across the tank from both devices and the varactor itself. We can go around the issue of tuning with different approaches as we will now show.

Consider first a single varactor solution. Assume a self-resonance frequency of 2.5 GHz for the inductor (or 0.4 pF of self-capacitance). As a rule of thumb, the varactor parasitic capacitance is 40-50% of the total tunable capacitance of the varactor. This should be at most 0.8 pF (assuming that all the remaining 1.6 pF are used for a single varactor). Another rule of thumb is that in any CMOS technology, the gate parasitic capacitance of a minimum-feature MOS transistor is given by

$$C_g \approx 1.5 \text{ fF} \times W(\mu\text{m}). \tag{24}$$

In other words, a transistor with a minimum channel length in any process has a gate capacitance of 1.5 fF per 1 μm of channel width. Note that this is the maximum varactor capacitance and is achieved in strong inversion.

This means that the NMOS varactor device width is around 500 μm for a maximum tuning range. The capacitance in (24) corresponds to the maximum NMOS varactor capacitance. The minimum capacitance is 2-3 times smaller. In this example, the minimum varactor capacitance is therefore around 300 fF. Therefore, the maximum frequency that this oscillator can reach is below 1.3 GHz, which can only be achieved when the varactor capacitance is at minimum for the entire oscillation cycle†. More information about the precise calculation of varactor tunability and the resulting oscillator sensitivity, are given in a later chapter about the role of the varactor in tuning and noise. For the moment, we can say that a single varactor at a limited oscillator swing can result in large phase noise performance degradation.

† The extremes of the frequency range are achievable under certain amplitudes and supply voltages. This topic is further discussed in the varactor chapter.

One other alternative is to limit the size of the varactor and add switchable fixed capacitance. This solution allows for the same tuning range only with lower VCO sensitivity as we discussed earlier.

The remaining element is the noise-filter. The filter capacitance can be as large as possible. As for the inductor, we select an inductor that is large enough to provide large impedance around $2f_0$. The inductor should be comparable in size to the main oscillator inductor with the highest Q. There are no unique component values for the noise filter. The three considerations in the design are:

1. filter corner frequency (ω_f): should be set below ~1/10 of the oscillation frequency.
2. filter impedance: inductor must be comparable to the main tank inductor.
3. inductor noise: Q must be high enough so that the inductor's thermal noise is much smaller than the current source thermal noise.

Finally, the device sizes for the transistors are to be selected. To guarantee startup, the loop gain has to exceed 1 and in practice should be 2-3. The tank impedance is equal to $Q\omega_0 L$, which equals 553 Ω. Therefore g_m of the switches has to exceed 5 mS. Consequently, the device size can be calculated:

$$\frac{W}{L} = \frac{g_m^2}{2\mu c'_{ox}\frac{I_o}{2}} \tag{25}$$

As for the current source, in the presence of a noise filter, the filter inductor allows the common source point of the FET switches to go below the drain of the current source. However, it is still a good practice to reduce the transconductance of the current source FET. The single-ended swing is limited to 0.65 V_{DD} or 1.625 V. Therefore the current source device, without a noise filter, could have a V_{DS} as low as 0.875 V. Assuming a threshold voltage of about 0.6 V, the gate-source voltage of the current source should not exceed 1.5 V. Using (25), the device ratio can be calculated. For reasons that will be clear when we discuss flicker noise, we choose a large channel length, much larger than the feature size, for example, 2 μm. Our oscillator design is now complete.

FIGURE 35 *Complete VCO schematic.*

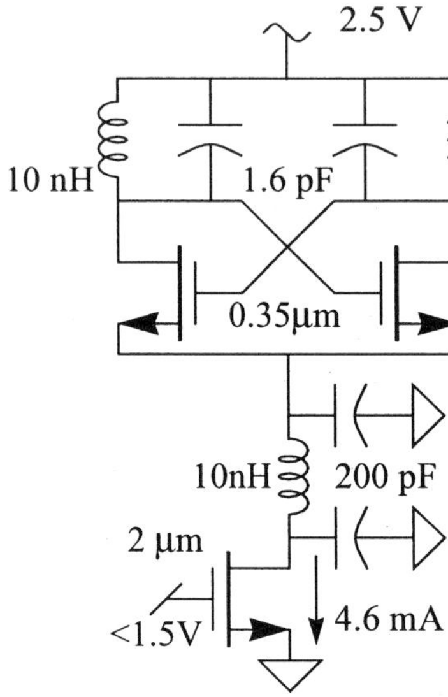

13 Anatomy of the Figure-of-Merit

The FOM given in (1) can be re-written in terms of circuit parameters using the phase noise model presented earlier,

$$\mathrm{FOM} = \frac{\omega_o^2}{\omega^2}\frac{1}{\frac{1}{V_o^2}\frac{kT}{C}\frac{\omega_o}{Q}\frac{2}{\omega^2}F}\frac{1}{I_o V_{DD}}. \tag{26}$$

Assume the oscillation amplitude is related to the supply voltage by a factor α.

$$V_o = \alpha V_{DD},\ \alpha \le 2 \tag{27}$$

The differential oscillation amplitude is related to the bias current as follows:

$$V_o = n\eta I_o Q\omega_o L \tag{28}$$

where L is the single-ended tank inductance.

Substituting from (27) and (28) into (26) we obtain the following simplified expression for FOM:

$$\mathrm{FOM} = \frac{\eta\alpha Q^2}{2kTF}. \tag{29}$$

Notice that the FOM is not a function of the inductance or the capacitance of the tank. However, it is a quadratic function of the tank quality.

To compare the relative merits of various oscillator topologies, we suggest the elimination of the dependence of FOM on Q. The new Topological FOM, TFOM is given by

$$\mathrm{TFOM} = \frac{\omega_o^2}{\omega^2 Q^2} \frac{1}{P\mathcal{L}(\omega)}. \tag{30}$$

Using the former substitutions, TFOM can be reduced to:

$$\mathrm{TFOM} = \frac{\eta\alpha}{2kTF} \tag{31}$$

Furthermore, to compare the merits of the topologies without considering the specifics of the design, or the designer's skill for that matter, we set α to the maximum value for all topologies. This simply means that we bias the oscillator for maximum amplitude disregarding reliability considerations. It is worth noting that if α is below the maximum value given in Table 1 then TFOM is no longer independent of the tank Q. To illustrate, it is possible to rewrite (31) as

$$\mathrm{TFOM} = \frac{\eta^2 Q\omega_o L I_o}{2kTFV_{DD}}, \tag{32}$$

which is directly derivable from (31) using (27) and (28).

In other words, TFOM given by (31) is the maximum "available" FOM per unit of Q. The designer may elect to design a suboptimal oscillator for reliability or power consumption as we saw in the GSM redesign example.

TABLE 1 *Maximum TFOM for current-biased VCO.*

Topology	α_{max}	η	$TFOM_{max}$
Tail-biased w/ filter	2	$2/\pi$	$\dfrac{2}{\pi kT(1+\gamma)}$
Complementary differential w/ filter	1	$4/\pi$	$\dfrac{2}{\pi kT(1+2\gamma)}$

From Table 1 it appears that the maximum TFOM for the tail/top biased VCO with a noise filter is higher than that of the complementary differential oscillator with a noise filter. The difference is about 2 dB if γ is equal to 4/3. This seems conflicting with what we stated in Section 11.4 about the advantages of the noise-filtered complementary differential topology.

To clarify, consider the other definition of TFOM given in (32), for two oscillators with the same current consumption, supply voltage, frequency and tank inductance, the noise-filtered complementary differential VCO has always a higher TFOM until the point where its amplitude, or a, is at maximum. Beyond that point, the regular tail-biased VCO with a noise filter has a better TFOM. This is the reason that the best-published VCO performance is always achieved using a tail-biased VCO rather than the complementary differential. However, for reliability concerns, the VCO is rarely operated in the very large amplitude region. Therefore, the complementary differential topology with a noise filter presents the best reliable topology.

References

[1] F. Svelto, S. Deantoni, and R. Castello, "A 1 mA, –120.5 dBc/Hz at 600 kHz from 1.9 GHz fully tunable LC CMOS VCO," in Custom IC Conf., Orlando, FL, pp. 577-580, 2000.

[2] M. A. Margarit, T. J. Leong, R. G. Meyer, and M. J. Deen, "A low-noise, low-power VCO with automatic amplitude control for wireless applications," *IEEE J. of Solid-State Circuits*, vol. 34, no. 6, pp. 761-771, 1999.

[3] H. Wang, "A 50 GHz VCO in 0.25m CMOS," in Intl. Solid-State Circuits Conf., San Francisco, CA, pp. 372-373, 2001.

[4] A. Dec and K. Suyama, "A 1.9-GHz CMOS VCO with micromachined electromechanically tunable capacitors," *IEEE J. of Solid-State Circuits,* vol. 35, no. 8, pp. 1231-1237, 2000.

[5] P. Kinget, "Integrated GHz voltage controlled oscillators," in *Analog Circuit Design: (X)DSL and other Communication Systems; RF MOST Models; Integrated Filters and Oscillators*, W. Sansen, J. Huijsing, and R. van de Plassche, Eds. Boston: Kluwer, 1999, pp. 353-381.

[6] Janusz Groszkowski, "The interdependence of frequency variation and harmonic content, and the problem of constant frequency oscillators," *Proc. of the IRE*, vol. 21, no. 7, pp.958-981, July 1933.

[7] HongMo Wang, "A solution for minimizing phase noise in low-power resonator-based oscillators," *Proc. of Int. Symp. on Circuits and Systems Conference*, ISCAS, vol. 3, pp. 53-56, Geneva, 2000.

[8] Kartikeya Mayaram, "Output voltage analysis for the MOS Colpitts oscillator," *IEEE Trans. on Circuits and Systems* Part-I Fundamental Theory and Applications, vol. 47, no. 2, February 2000.

[9] R. Aparicio, and A. Hajimiri, "A noise-shifting differential Colpitts VCO," *IEEE J. of Solid-State Circuits,* vol. 37, no. 12, pp. 1728-1736, 2002.

[10] Craninckx, J.; Steyaert, M.; Miyakawa, H., "A fully integrated spiral-LC CMOS VCO set with prescaler for GSM and DCS-1800 systems," *Proc. of the IEEE Custom Integrated Circuits Conference*, pp. 403-406, Santa Clara, 1997.

[11] J. E. Post, Jr., I. R. Linscott, and M. H. Oslick, "Waveform symmetry properties and phase noise in oscillators," *Electron. Lett.*, vol. 34, no. 16, pp. 1547-1548, Aug. 1998.

[12] B. De Muer, M. Borremans, M. Steyaert, and G. Li Puma, "A 2-GHz low-phase-noise integrated LC-VCO set with flicker-noise upconversion minimization," *IEEE J. of Solid-State Circuits,* vol. 35, no. 7, pp. 1034-1038, 2000.

[13] A. Hajimiri, and T. H. Lee, "Design issues in CMOS differential LC oscillators," *IEEE Journal of Solid-State Circuits,* vol. 34 no. 5, pp. 717 -724 May 1999.

[14] K. Hoshino, E. Hegazi, J. Rael and A.A. Abidi, "A 1.5 V, 1.7 mA 700 MHz CMOS LC oscillator with no upconverted flicker noise," *Proc. of the European Solid State Circuits Conference*, Villach, 2001.

[15] A. A. Abidi, "High frequency noise in FETs with small Dimensions," *IEEE Trans. on Electron Devices*, vol. ED-33, no. 11, pp. 1801-1805, 1986.

[16] J. Lee, A. A. Abidi, and N. G. Alexopoulos, "Design of spiral inductors on silicon substrates with a fast simulator," in *European Solid-State Circuits Conf.,* The Hague, The Netherlands, pp. 328-331, 1998.

[17] C. P. Yue and S. S. Wong, "On-chip spiral inductors with patterned ground shields for Si-based RF ICs," *IEEE J. of Solid-State Circuits,* pp. 743-752, 1998.

[18] H. Darabi and A. Abidi, "Noise in CMOS mixers: a simple physical model," IEEE Journal of Solid State Circuits, vol. 35, no. 1, pp. 15-25, 2000.

[19] J. Phillips and K. Kundert, "Noise in mixers, oscillators, samplers, and logic: an introduction to cyclostationary noise," in Custom IC Conf., Orlando, FL, pp. 431-438, 2000. Also available from *www.designers-guide.com/Theory.*

[20] E. Hegazi, H. Sjoland and A. A. Abidi, "A filtering technique to lower LC oscillator phase noise," *IEEE Journal of solid-state circuits,* Vol. 36 no. 12, pp-1921-1930 Dec. 2001.

[21] A. Kral, F. Behbahani, and A. A. Abidi, "RF CMOS oscillators with switched tuning," *Proc. IEEE Custom Integrated Circuits Conf.*, Santa Clara, pp. 555-558, 1998.

[22] J. J. Rael and A. A. Abidi, "Physical processes of phase noise in differential LC oscillators," in *Proc. of the IEEE Custom Integrated Circuits Conference,* pp. 569-572, Orlando, 2000.

[23] W. P. Robins, *Phase Noise in Signal Sources (Theory and Applications)*, Peter Peregrinus, London, 1982.

[24] A. A. Abidi, "How phase noise appears in oscillators," in Analog Circuit Design: RF A/D Converters, Sensor and Actuator Interfaces, Low-Noise Oscillators, PLLs, and Synthesizers, R. J. van de Plassche, J. H. Huijsing, and W. Sansen, Eds. Boston: Kluwer, 1997, pp. 428-448.

[25] A. Hajimiri, and T. H. Lee, "A general theory for phase noise in electrical oscillators," *IEEE Journal of Solid-State Circuits*, vol. 33, no. 2, pp. 179-194, Feb. 1998.

[26] A. Rofougaran, G. Chang, J. J. Rael, J. Y.-C. Chang, M. Rofougaran, P. J. Chang, M. Djafari, M. K. Ku, E. Roth, A. A. Abidi, and H. Samueli, "A single-chip 900 MHz spread-spectrum wireless transceiver in 1μm CMOS (Part I: architecture and transmitter design)," *IEEE J. of Solid-State Circuits,* vol. 33, no. 4, pp. 515-534, 1998.

[27] N. Itoh, B. De Muer, and M. Steyaert, "Low supply voltage fully integrated CMOS VCO with three terminals spiral inductor," *in European Solid-State Circuits Conf.*, Duisburg, Germany, pp. 194-197, 1999.

[28] Alan Hastings, *The Art of Analog Layout*, Prentice Hall, 2001.

[29] C. Samori, A. L. Lacaita, F. Villa, and F. Zappa, "Spectrum folding and phase noise in LC tuned oscillators," *IEEE Trans on Circuits and Systems II: Analog and Digital Signal Processing*, vol. 45, no. 7, pp. 781-90, 1998.

[30] F. Behbahani, W. Tan, A. Karimi-Sanjaani, A. Roithmeier, and A. A. Abidi, "A broad-band tunable CMOS channel-select filter for a low-IF wireless receiver," *IEEE J. of Solid-State Circuits,* vol. 35, no. 4, pp. 476-489, 2000.

[31] Tchamov and P. Jarske, "New low-power GHz-range resonance-ring ICO/ VCO," in *Int. Conf. on Electronics, Circuits, and Systems,* Rodos, Greece, pp. 358-361, 1996.

6 Flicker Noise

1 Flicker Induced Phase Noise

Approaching the carrier, the slope of the phase noise spectrum in CMOS VCOs increases from –20 to –30 dB/decade. This is ascribed to the up conversion of flicker noise in FETs. The analysis presented in the previous chapter is reviewed to see if it explains this up conversion.

Flicker noise in the tail current source at a low frequency of ω_m indeed up converts to $\omega_0 \pm \omega_m$ and enters the resonator, but as AM, not PM noise. Therefore, in the absence of a high gain varactor to convert AM to FM [1], flicker noise in the tail current will not appear as phase noise. Next, consider flicker noise in the differential pair. The preceding analysis says that this modulates zero crossings, and injects a noise current into the resonator consisting of flicker noise sampled by an impulse train with frequency $2\omega_0$. Thus, noise originating at frequency ω_m produces currents at ω_m and at $2\omega_0 \pm \omega_m$. Both frequencies are strongly attenuated by the resonator, and neither explains flicker-induced phase noise at $\omega_0 \pm \omega_m$.

Therefore, the model presented so far predicts that none of the noise at low frequencies produces phase noise around the carrier. Yet measurements show that close to the carrier, there is up converted flicker noise, therefore there must be another mechanism.

2 FM Due to Modulated Frequency Content

Figure 1a shows a model of the oscillator where it is normally assumed that the negative resistance produces a sinusoidal current to compensate for the loss in the resonator. However, because the negative resistance is typically implemented with a non-linear circuit, this current is more like a square-wave, which is rich in harmonics, Figure 1b. Normally, these harmonics are neglected, but they must flow somewhere in the circuit and in fact at these high frequencies, the capacitor offers the lowest impedance path.

FIGURE 1 *The output current from the active circuit is a square wave, not a sine wave. The fundamental component flows into the loss resistance, while the high frequency harmonics flow into the capacitor, resulting in a shift between the zero crossings of the output voltage and the output current.*

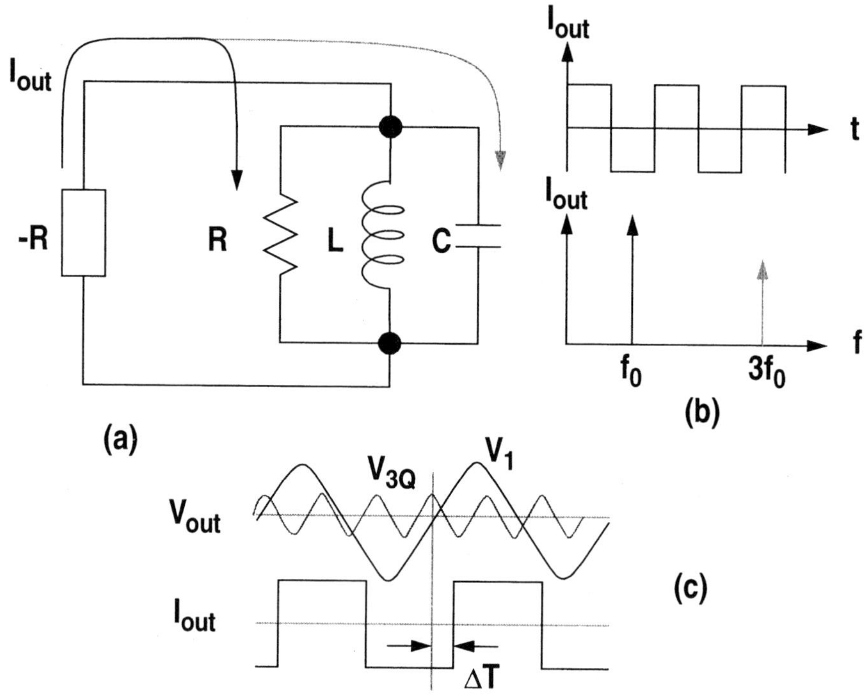

As a result, the phase relationship between the zero crossings of the current and the voltage is upset, as shown in Figure 1c. This shift in phase due to harmonics produces a change in oscillation frequency given by (1), which was predicted by Groszkowski in 1933 [2].

$$\frac{\omega_m}{\omega_0} = \frac{1}{2Q^2} \sum_{n=2}^{\infty} \frac{n^2(1-n^2)}{(1-n^2)^2 + n^2/Q^2} m_n^2 \quad (1)$$

The normalized change in frequency depends on the harmonic levels, m_n, and the quality factor, Q, of the resonator. Now if noise *modulates* the harmonic level, it will cause frequency modulation and *that* will be a mechanism of phase noise.

In the LC oscillator, the amount of harmonic content is a function of the bias current, I_0. A larger current produces a larger amplitude (relative to the transition region of the differential pair), which generates a current waveform that is switched more quickly and contains a higher harmonic content. The sensitivity $\partial\omega/\partial I_0$ is responsible for "indirect" FM [3] due to flicker noise in I_0.

While the concept of the oscillation frequency being a function of amplitude may seem such a minor effect that it can be ignored, it is often used in the design of precision frequency references such as crystal oscillators for watches. In such a watch, an error as minor as 20 ppm in the oscillation frequency results in a loss of 10 minutes in a year's time. Often, amplitude control loops with very slow response times are used in watch applications, allowing the minimum amount of excess gain to be used and still insure that the circuit will start oscillating. Vittoz describes the use of an AGC circuit in a crystal oscillator to "reduce distortions" since nonlinearities have a "devastating effect on frequency stability" [4]. His paper also includes a graphical method to determine the frequency of oscillation.

These ideas can be found elsewhere as well. Gavra and Ermolenko, in a Russian journal, describe how "for high-frequency stability, it is necessary to stabilize the amplitude of oscillation at the lowest possible level" [5]. From these examples, it is clear that these are not obscure or useless ideas.

Typically, a frequency offset due to this effect is not as important in low cost mobile applications because low cost crystals are used with inherent frequency errors up to 20 ppm. It is possible to use such poor references because the time period of interest is so short (i.e. the length of a data packet).

2.1 Groszkowski in a van der Pol Oscillator

Next, consider the use of a third order non-linear g_m circuit, as shown in Figure 2. The analysis is simplified by only considering energy at f_0 and $3f_0$. While this analysis is not thorough, it is accurate for circuits with only a small amount of non-linearity and provides intuition into the operation of the circuit. A more thorough analysis would include more frequency terms. This type of analysis is very similar to what occurs within a harmonic-balance type simulator, where the same errors occur if only a few harmonics are used.

FIGURE 2 *A third-order, non-linearity in the active circuit is considered in this oscillator.*

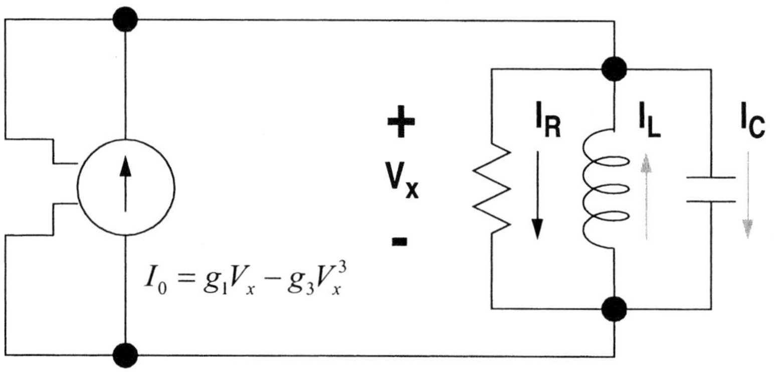

Typically the active circuit is designed to over-compensate for the loss in the resonator, ensuring the oscillator will self-start on noise in the circuit. The product $g_1 R$ is referred to as the small signal loop gain and is typically larger than unity to ensure "self-start" operation. As the amplitude increases, the third order nonlinearity reduces the amount of compensation current that is injected into the resonator. Steady-state occurs when the amplitude of oscillation generates a current that exactly equals the current flowing through the resistor. However, this built in amplitude control does not come without cost. The nonlinearity also generates a current at the third harmonic, which flows through the capacitor, producing a quadrature voltage at $3f_0$. This harmonic mixes with the fundamental in the active circuit and generates a quadrature current at the fundamental.

$$V_{\text{out}} \cong V_1 \cos(\omega t) + V_{3\text{Q}} \sin(3\omega t) \tag{2}$$

$$I_{\text{out}} \cong \left(g_1 V_1 - \frac{3}{4} g_3 V_1^3\right) \cos(\omega t) - \frac{3}{4} g_3 V_1^2 V_{3\text{Q}} \sin(\omega t) \tag{3}$$

One result of the quadrature current is a phase shift between the output voltage and the active circuit current. This phase shift is simply the inverse tangent of the ratio of the quadrature and in-phase component.

$$\phi = \text{atanh}\left[\frac{-3 V_1 V_{3\text{Q}}}{4\frac{g_1}{g_3} - 3 V_1^2}\right] \tag{4}$$

Previously, all the current from the active circuit flowed into the resistor. Since the resistor current must be in-phase, the quadrature component must flow into the resonator. When KCL is performed, the quadrature component contains a new term and is given in (5).

$$\frac{3}{4} g_3 V_1^2 V_{3\text{Q}} = V_1 C\omega - \frac{V_1}{L\omega} \tag{5}$$

This results in oscillation no longer occurring at ω_0 but rather at a slightly lower frequency. Figure 3 shows the calculated frequency of oscillation as well as the simulated frequency of oscillation. This is a simplified version of the rigorous analysis performed by Groszkowski and given in (1) [2].

This quadrature current can be modeled with a “Groszkowski” capacitance as given in (6). This expression was simplified by approximating ω with ω_0. For small frequency deviations, the error is acceptable. This Groszkowski capacitance is added to the tank capacitance resulting in a total capacitance that accurately predicts the frequency of oscillation.

$$C_{\text{Gr}} \approx \frac{3}{4} \frac{V_1 V_{3\text{Q}}}{\omega_0} g_3 \tag{6}$$

FIGURE 3 *Frequency of oscillation versus small signal gain.*

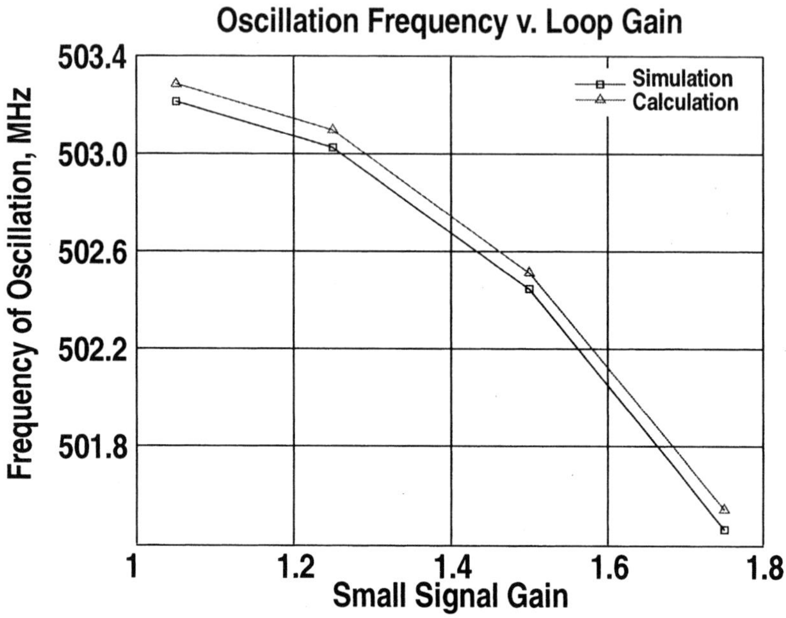

2.2 Groszkowski in the Differential Pair Oscillator

While Groszkowski's model of the oscillator is accurate, the derivation is complicated and is not intuitive. Further, to use it in the context of phase noise, an expression for the sensitivity of the frequency shift as a function of harmonic content is required. In this section, a simplified model is derived using the same methods previously developed.

First, the cross-coupled differential pair is modeled as a hyperbolic tangent function.

$$i_{out} = -I_0 \tanh\left(\frac{v_{out}}{V_w}\right) \tag{7}$$

Where I_0 is the tail current and V_w is approximately the "linear" region of the I-V characteristic. The output current in these oscillators varies from a sine wave to nearly a square wave depending on the excess gain in the loop. The current consists of a fundamental component, a 3rd harmonic, and higher harmonics, which are ignored.

Next, assume the voltage across the tank, v_{out}, consists of the following terms:

$$v_{out} = V_1 \sin\omega t + V_{3I}\sin 3\omega t + V_{3Q}\cos 3\omega t\,. \tag{8}$$

The 3rd harmonic in v_{out} arises from the 3rd harmonic of current flowing mostly through the capacitor. In the extreme case, the output current is a square wave resulting in the following values for the voltage terms:

$$V_1 \cong \frac{2}{\pi} I_0 R \tag{9}$$

$$V_{3I} \cong 0 \tag{10}$$

$$V_{3Q} \cong \frac{2}{3\pi}\frac{I_0}{3\omega C} \tag{11}$$

Proving that even in the extreme case of high excess loop gain, the most nonlinear case, the third harmonic voltage is much less than the fundamental.

$$V_{3Q} = \frac{V_1}{9Q} \ll V_1 \tag{12}$$

This fact allows (7) to be expanded into a Taylor series. The first few terms are given below.

$$i_{out} = -I_0 \tanh\left(\frac{V_1 \sin\omega t + V_{3Q}\cos 3\omega t}{V_w}\right) \tag{13}$$

$$i_{out} \cong -I_0 \tanh\left(\frac{V_1 \sin\omega t}{V_w}\right) - I_0 \frac{V_{3Q}\cos 3\omega t}{V_w} + I_0 \frac{V_{3Q}\cos 3\omega t}{V_w}\tanh^2\left(\frac{V_1 \sin\omega t}{V_w}\right) \tag{14}$$

The first term corresponds to the *disturbance-free* square-wave form. The second term is a third harmonic tone that is in quadrature with the waveform. The third term is a mixing term that operates on the third harmonic term. Next, consider two cases: low loop gain which operates at critical oscillation and high loop gain which results in hard limiting.

2.3 Critical Oscillation

In the first case, the amplitude of the fundamental, V_1, is comparable to the transition region of the *tanh* function, Figure 4a. The negative of the mixing waveform is shown in Figure 4b. The waveform bares a strong resemblance to a full-wave rectified sine wave and will be approximated as such. The Fourier series for a full-wave rectified sine wave is given in (15).

$$p(t) = \frac{4}{\pi}\left(\frac{1}{2} + \frac{1}{1\cdot 3}\cos 2\omega t + \frac{1}{3\cdot 5}\cos 4\omega t + \ldots\right) \tag{15}$$

FIGURE 4 At critical oscillation, the V_1 component of v_{out} is comparable to the transition region, V_w, of the oscillator. Other than an inversion and a DC offset, the resulting waveform is roughly the same as full-wave rectified sine wave.

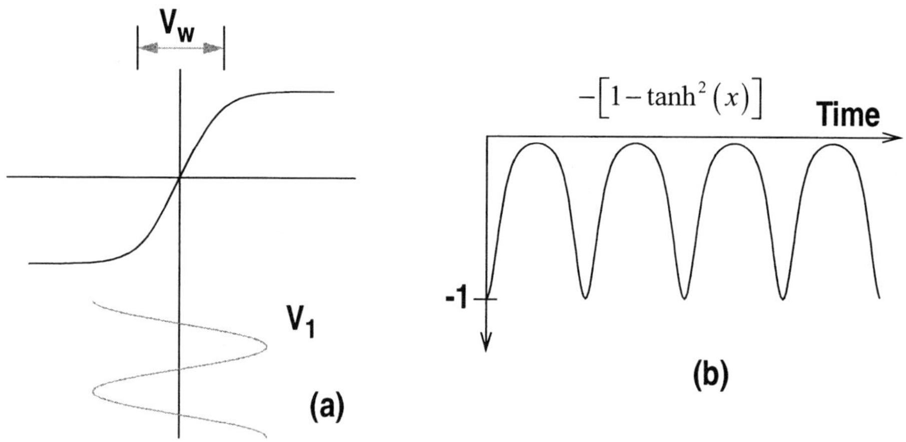

The mixing action between the second harmonic of (15) and the third harmonic of the waveform produces a term at the fundamental frequency:

$$\vec{I} = I_0\frac{V_{3Q}}{V_w}\cos 3\omega t\frac{4}{3\pi}\cos 2\omega t = \frac{2}{3\pi}I_0\frac{V_{3Q}}{V_w}\cos\omega t + \ldots \tag{16}$$

This current component is in quadrature with the fundamental and looks capacitive relative to the fundamental of the tank voltage. The effective capacitance is:

$$C_{Gr} = \frac{\vec{I}}{\omega \vec{V}} = \frac{2}{3\pi} I_0 \frac{V_{3Q}}{\omega V_w V_1} \tag{17}$$

Qualitatively, as V_1 increases slightly and enters soft clipping, the 3rd harmonic rises dramatically. Thus, the capacitor changes sharply.

The equivalent circuit for the active circuit portion of the oscillator is given in Figure 5. As before, the active circuit produces a negative resistance that compensates for the loss in the tank. This analysis shows an additional *reactive* component that is dependent on the harmonic content of the waveforms. As the harmonic content changes, the effective capacitance changes, modulating the frequency of oscillation. This FM effect is a source of flicker induced phase noise.

FIGURE 5 *The equivalent impedance of the differential pair consists of a negative resistance and a modulated capacitance.*

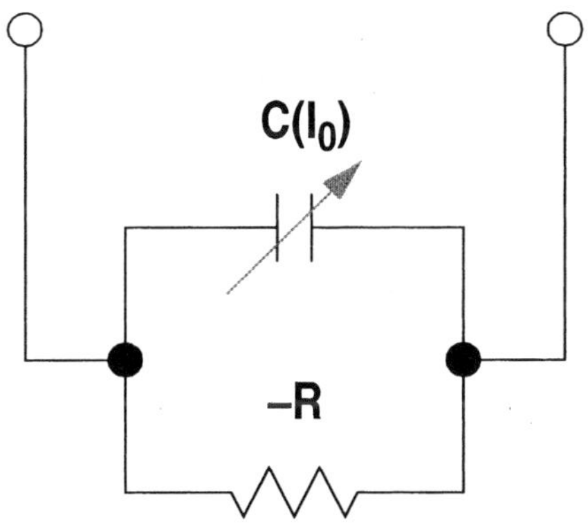

The equation above was developed by assuming a shape of the multiplying function and only looking at the amplitude of the signal that results when the 3rd harmonic of the output signal is multiplied by the tone of the multiplying function at twice the fundamental frequency. However, the multiplier function tone at four times the fundamental will also produce an output at the fundamental frequency. Further, other disturbance tones at the output voltage such as the 5th and 7th harmonic will produce an output at the fundamental. These contributions drop off in amplitude quickly and a close formed solution is easily derived. First, V_3 in (12) is modified.

$$V_{3Q} \cong \frac{2}{3\pi}\frac{I_0}{3\omega C} = \frac{2}{\pi}\frac{I_0}{3}\frac{1}{3\omega C} = \frac{V_k}{9} \tag{18}$$

Assume the relative level of the harmonics beyond the 3rd is equal to the ratio in the square wave case. The amplitude of the output at the fundamental is given below. The tones are grouped as harmonics of *tanh* function mixing with the harmonics of the full rectified sine wave, as in (15).

$$V_{3Q} = k\left[\frac{1}{3}\cdot\frac{1}{3}\left(\frac{1}{1\cdot 3}+\frac{1}{3\cdot 5}\right)+\frac{1}{5}\cdot\frac{1}{5}\left(\frac{1}{3\cdot 5}+\frac{1}{5\cdot 7}\right)+\ldots\right] \tag{19}$$

This can be generalized as:

$$V_{3Q} = k\left[\sum_{n=1}^{\infty}\frac{1}{(2n+1)^3(2n-1)}+\sum_{n=2}^{\infty}\frac{1}{(2n-1)^3(2n+1)}\right] \tag{20}$$

$$V_{3Q} = k\left[\sum_{n=1}^{\infty}\frac{1}{(2n+1)^3(2n-1)}+\sum_{n=1}^{\infty}\frac{1}{(2n-1)^3(2n+1)}-\frac{1}{3}\right] \tag{21}$$

$$V_{3Q} = k\left[\frac{2}{3}-\frac{\pi^2}{16}\right]. \tag{22}$$

In the single tone derivation, the value of (21) was $k/27$. Since each term contributes a quadrature current that sums to generate an effective capacitance, the equation in (17) can be scaled by the ratio between the two. The final equation for C_{eq} is given below.

$$C_{Gr} = \frac{2/3-\pi^2/16}{1/27}\frac{2}{3\pi}I_0\frac{V_{3Q}}{\omega V_w V_1} = \left[\frac{2}{3}-\frac{\pi^2}{16}\right]\frac{18}{\pi}I_0\frac{V_{3Q}}{\omega V_w V_1} \tag{23}$$

2.4 Hard Limiting

In this case, the excess loop gain forces the oscillator into hard limiting. Here, the waveforms now look entirely different than before, Figure 6. The multi-

FIGURE 6 *The multiplying waveform for the hard limiting case approaches an impulse train.*

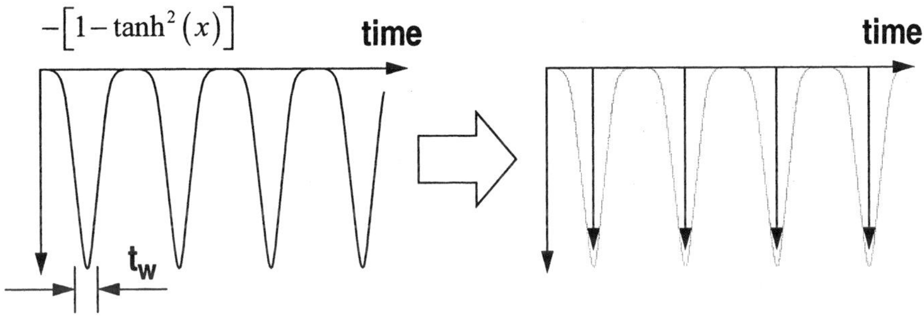

plying waveform has narrowed and approaches an impulse train. Each pulse is t_W wide. The train of pulses can be approximated by delta functions with the same area. Thus, in the time-domain the multiplying waveform is:

$$p(t) = \frac{4}{\pi}\omega t_w(1 + \cos 2\omega t + \cos 4\omega t + \ldots) \tag{24}$$

After multiplication with the third harmonic of the tank voltage, the second and third term of the Fourier series contribute a current at the fundamental frequency,

$$\vec{I} = I_0 \frac{V_{3Q}}{V_w} \cos 3\omega t \frac{4}{\pi}\omega t_w \cos 2\omega t = \frac{2}{\pi}\omega t_w I_0 \frac{V_{3Q}}{V_w} \cos \omega t + \ldots, \tag{25}$$

which defines a capacitance relative to the tank voltage. The effective capacitance is

$$C_{Gr} = \frac{\vec{I}}{\omega \vec{V}} = \frac{2}{\pi} I_0 \omega t_w \frac{V_{3Q}}{V_w \omega V_1} = \frac{2}{\pi} I_0 t_w \frac{V_{3Q}}{V_w V_1}. \tag{26}$$

In hard limiting, we can use the following relationships:

$$V_1 = \frac{2}{\pi} I_0 R = \frac{2}{\pi} I_0 \frac{Q}{\omega C} \tag{27}$$

$$t_w = \frac{V_w}{\omega V_1} \tag{28}$$

$$V_{3Q} = \frac{V_1}{9Q}, \tag{29}$$

where Q is the tank quality factor. Substituting into (26), the effective capacitance is:

$$C_{Gr} = \frac{4}{\pi} I_0 \frac{1}{9Q\omega V_1} = \frac{C}{9Q^2} \tag{30}$$

C is the fixed capacitance in the tank.

In the hard limiting case, the level of the harmonics does not change. Therefore, the Groszkowski effect in hard limiting leads to a fixed shift in oscillation frequency. From the above expression, the normalized frequency shift resulting from 3rd harmonic generation is:

$$\frac{\Delta\omega}{\omega} \cong -\frac{1}{2}\frac{C_{Gr}}{C} = -\frac{1}{2}\frac{1}{(3Q)^2} \tag{31}$$

This equation has the same format as (1).

Just as before, the expression for the equivalent capacitance can be expanded to include more harmonics. Again, using the definition in (25), the total sum is given below.

$$\vec{I} = k\omega t_w \left[\frac{1}{3} \cdot \frac{1}{3}(1+1) + \frac{1}{5} \cdot \frac{1}{5}(1+1) + \dots\right] \tag{32}$$

$$\vec{I} = 2k\omega t_w \left[\sum_{n=1}^{\infty} \frac{1}{(2n+1)^2}\right] \tag{33}$$

$$\vec{I} = 2k\omega t_{\mathrm{w}} \frac{\pi^2 - 8}{8} \tag{34}$$

In the initial analysis, the second term in (32) was equal to 1/9. A more accurate model for the capacitance is derived by scaling (30) by the ratio. The final equation for C_{eq} is given below.

$$C_{\mathrm{Gr}} = \frac{9(\pi^2 - 8)}{8} = \frac{\pi^2 - 8}{8} \frac{C}{Q^2} \tag{35}$$

To verify these equations, the oscillator is simulated with a simple macro model where the nonlinearity is captured by a hyperbolic tangent, Figure 7a. SpectreRF is used to simulate the effective capacitance of the *tanh* function.

FIGURE 7 *A simple oscillator is used to verify the Groszkowski effect.*

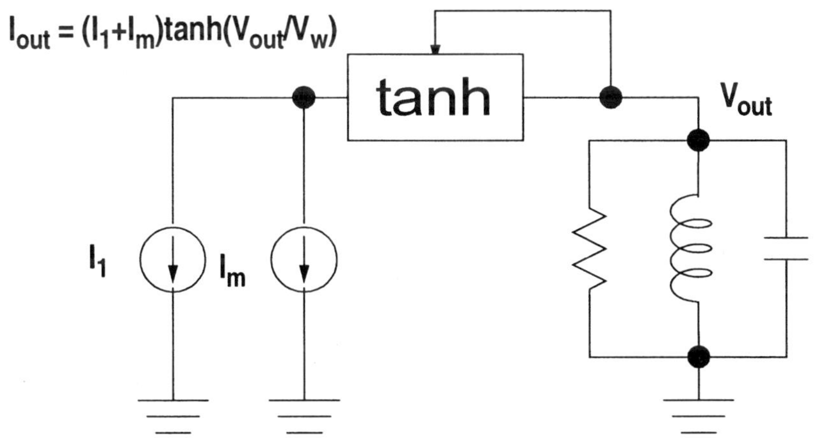

This is plotted as a function of gain, which is defined as the inverse of V_{w}. The multi-tone equations predict the simulation very well, (8).

3 Switch Voltage Noise Modulates Capacitance

Groszkowski is not the only mechanism of indirect FM. Flicker noise in the differential pair also appears around the carrier but through a different process. Consider the oscillation at the two nodes V_{p} and V_{n} shown in Figure 9b.

FIGURE 8 *Effective Groszkowski capacitance versus gain ($1/V_w$). Excellent agreement between simulation results and equations (23) and (35).*

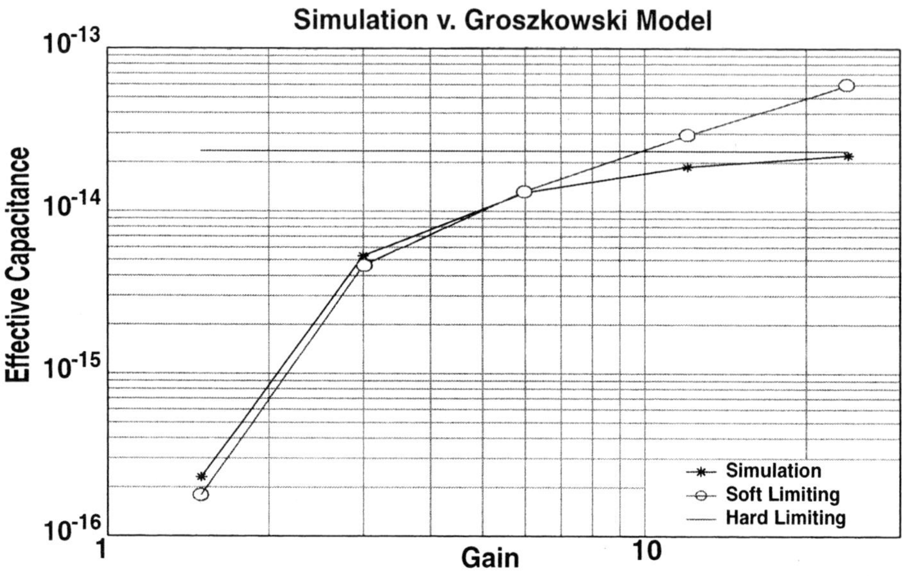

Assume a differential voltage across the tank that is given by (36) and is shown in Figure 9b.

$$v_o = V_1 \sin(\omega_0 t) \tag{36}$$

This oscillation will produce a voltage waveform at the tail at twice the oscillation frequency. Next, suppose the noise in this differential pair is modeled with a single noise source. When the right transistor is on, the tail voltage rises to a nominal value and goes down Figure 9c. On the other half cycle, when the left transistor turns on, the tail voltage rises to a noisy value. This results in a common-mode voltage that can be approximated as in Figure 9c. The minimum occurs when the differential output voltage is equal to zero while the maximum voltage occurs just when one side enters the linear region. This maximum occurs at time t_l in Figure 9c. The value of t_1 is given in (37).

$$t_1 = \frac{T_0}{2\pi}\sin^{-1}\left(\frac{V_T}{V_0}\right) \approx \frac{T_0}{2\pi}\frac{V_T}{V_0} \tag{37}$$

FIGURE 9 *Flicker noise in switching devices results in equivalent fluctuating negative capacitance.*

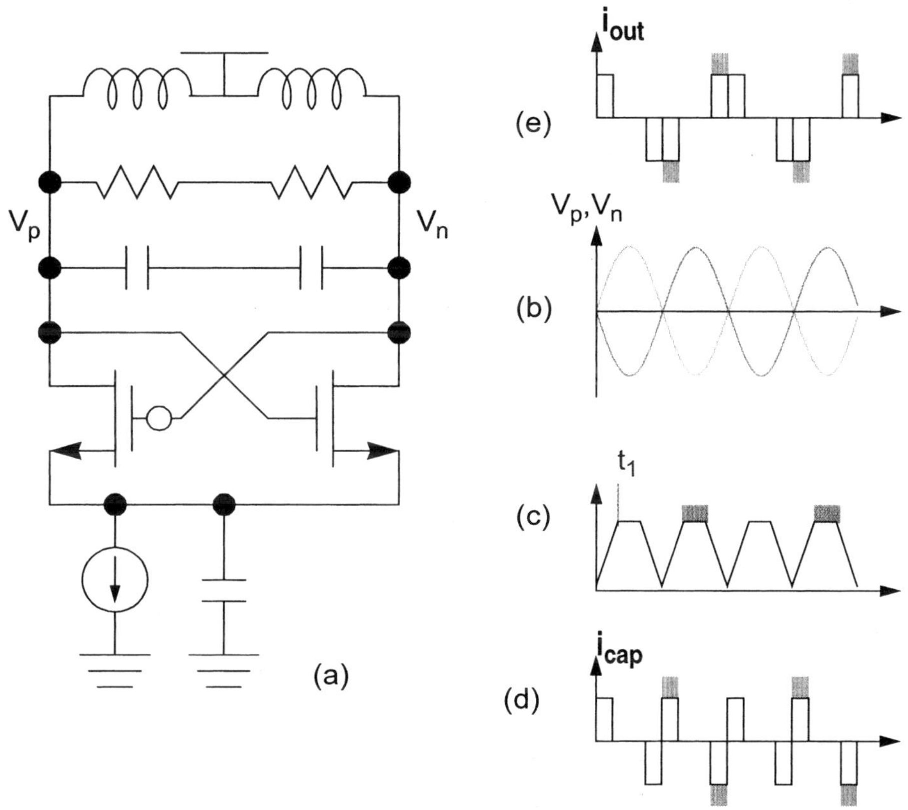

If the noise is modeled as an offset voltage, simulations show the common-mode does not rise to the full value but instead shifts the whole common-mode up by half the voltage and affected half rises by an addition half. So the height of the gray region of Figure 9c is simply $v_{os}/2$.

The waveform at the common-mode point induces a current in the tail capacitor, Figure 9d, which is simply the time derivative of the waveform in Figure 9c. Since the time difference is known and the amplitude is known, the height of Figure 9d is given in (38).

$$\Delta i_{\text{cap}} = C\frac{v_{\text{os}}/2}{t_1} \tag{38}$$

Now, the capacitor current flows through the cross-coupled switching differential pair and is commutated at the output before injection into the resonator, as shown in Figure 9e. The fundamental component of this commutated current describes a phase relationship with the voltage, which corresponds to a negative capacitance.

$$i_{\text{diff}} = -C\frac{d}{dt}v_{\text{out}} \tag{39}$$

This value is easily found by calculating the quadrature component of the gray squares in Figure 9e; include a half term for the differential component.

$$i_{1,q} = \frac{2}{T}\int_0^T \Delta i_{\text{out}}(t)\cos(\omega t)dt \tag{40}$$

$$i_{1,q} = \frac{2}{T}\frac{C_t v_{\text{os}}/2}{t_1}\left[-\int_{T/2}^{T/2+t_1}\cos(\omega t)dt + \int_{T-t_1}^{T}\cos(\omega t)dt\right] \tag{41}$$

$$i_{1,q} = \frac{v_{\text{os}}}{t_1}\frac{C_t}{\pi}\frac{V_{\text{T}}}{V_1} \tag{42}$$

As a result, the effective capacitance is a function of the noise. Therefore, the differential pair presents to the resonator a negative resistance and a fluctuating capacitance due to the capacitor and the noise. This fluctuating capacitor causes frequency modulation within the resonator and up-converts flicker noise around the carrier.

The capacitance is simply the quadrature current in (40) scaled by the output voltage and oscillation frequency.

$$C_{\text{eff}} = \frac{i_{1,q}}{V_1\omega} \tag{43}$$

$$C_{\text{eff}} = \frac{v_{\text{os}}}{\omega t_1}\frac{C_t}{\pi}\frac{V_{\text{T}}}{V_1^2} \tag{44}$$

$$C_{\text{eff}} = \frac{C_t v_{\text{os}}}{\pi V_1} \frac{V_T/V_1}{\operatorname{asin}(V_T/V_1)} \tag{45}$$

The effective capacitance is proportional to the tail capacitance and is strongly attenuated by the amplitude of oscillation. In actual designs, the tail capacitance should be minimized to reduce this source of modulation.

Even though drastic simplifications were made, these equations still accurately predict the quadrature current and the effective capacitance. Table 1 compares (40) and (43) versus simulation. The parameters of the simulation are also given in Table 1.

TABLE 1 *Comparison of calculated and simulated values. Parameters:* $V_1 = 882mV$, $C_t = 50f$, $V_T = 0.5$

Parameter	Calculated	Simulated v_{os} = +10mV/–10mV
Quadrature Current, i_{1q}	475.2nA	–478.5nA/384.6nA
Capacitance, C_{eff}	0.1704fF	–0.1729fF/0.1390fF

This simplified model is easily expanded to take into account the C_{gs} capacitance of the switching devices. The C_{gs} capacitors are lumped with the tail capacitance, (46).

$$C_{\text{tot}} = C_t + \frac{C_{gs}}{2} \tag{46}$$

The factor of 1/2 in the C_{gs} term comes from the fact that together, the C_{gs} capacitors form a series path. This was also simulated and matched very well.

4 Frequency Modulation by the Current Source

The next mechanism for flicker induced phase noise is similar to the Groszkowski effect described in Section 2. However, it only appears when parasitic capacitances are present at the tail and across the C_{gs} of the differential pair devices, Figure 10. In the theoretical case where these capacitances are zero, only the Groszkowski effect produces frequency modulation due to the current source.

FIGURE 10 *FM due to flicker noise in current source.*

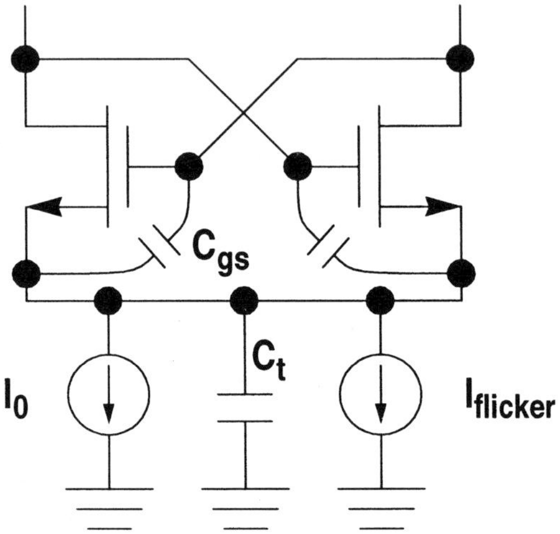

First some basic frequency modulation theory is reviewed. The bias current source is decomposed into a noise-free bias component and an equivalent flicker noise component, (47).

$$I_0 = I_{bias} + I_{flicker} \tag{47}$$

The change in oscillation frequency that occurs due to a change in the bias current is modeled as a bias dependent capacitance that appears in parallel with the tank. This concept was shown previously in Figure 3 in Chapter 3. If the bias dependent capacitance is linearized at the operating point, it can be modeled as (48).

$$C_m = k\Delta i \tag{48}$$

The frequency of oscillation with this modulated capacitance is easily calculated, (49).

$$\omega_{osc} = \frac{1}{\sqrt{L(C + C_m)}} \tag{49}$$

$$\omega_{osc} \approx \omega_0 - \frac{\omega_0}{2C} k \Delta i \tag{50}$$

Next, the FM coefficient is defined, (51).

$$\frac{\Delta\omega}{\Delta i} = -\frac{\omega_0}{2C} k \tag{51}$$

Finally, the equivalent phase noise sidebands are calculated with FM theory.

$$\mathcal{L}(\omega_m) = \frac{1}{8}\left(\frac{\Delta\omega}{\Delta i}\right)^2 I^2_{\text{flicker,rms}} \tag{52}$$

Returning to the oscillator in Figure 10, the analysis of the equivalent capacitance is based on the varactor paper by Hegazi [1]. In this paper, Hegazi realized that the instantaneous capacitance seen by the tank is *not* constant but changes with time. Since the frequency of the varying capacitance is periodic with frequency $2\omega_0$, it can be described with a Fourier series:

$$C_{inst}(t) = \sum_{n=0}^{\infty} C_{2n} \cos(2n\omega_0 t) \tag{53}$$

After accounting for the harmonics in the output voltage and the mixing that occurs, Hegazi derived an effective capacitance that is seen by the oscillator and is given in (54).

$$C_{eff} = C_0 - \frac{1}{2} C_2 \tag{54}$$

This same idea is used to analyze how the effective capacitance of the oscillator changes as a function of bias current, capacitance at the tail, and the C_{gs} of the switching devices, Figure 10. During the course of the oscillation, the tank sees a time varying capacitance, as shown in Figure 11.

It is straight forward to derive equations for the capacitance at each point. In the balanced condition, both transistors act as simple transconductance stages. Since the circuit is truly differential at this point, the common-mode point is a virtual ground, shorting the tail capacitor leaving only C_{gs} in parallel with the resonator. The equivalent capacitance is given in Equation 55.

FIGURE 11 *The resonator sees a time varying capacitance. When in the balanced condition the capacitance is a positive value. As the oscillation amplitude increases, the capacitance becomes negative. Finally, at the peak of oscillation, the capacitance is positive.*

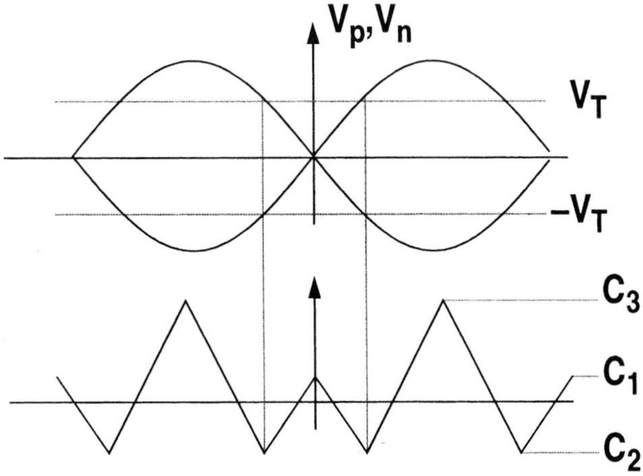

$$C_1 = C_{gs} \tag{55}$$

From the balanced condition, the differential voltage increase to the point that all the current is switch to one side. Here, one device is off and the other is modeled as a simple transconductance stage. The instantaneous capacitance at this point is given in (56).

$$Z_2 = \frac{2[g_m + s(2C_{gs} + C_t)]}{s[-g_m C_t + 2sC_{gs}(2C_{gs} + C_t)]} \tag{56}$$

By assuming small values for C_{gs} and C_t, this equation is simplified by ignoring terms that are small at the oscillation frequency.

$$Z_2 \approx \frac{-1}{sC_t/2} \tag{57}$$

This results in an instantaneous capacitance that is given in (58)

$$C_2 = \frac{-C_t}{2} \tag{58}$$

Finally, the last point occurs at the peak of the oscillation. Here all the current is switched to one device that is in triode while the other device is off. The triode region of a long channel device can be modeled as two back-to-back saturated device as shown in Figure 12a.

FIGURE 12 *(a) Triode device can be modeled with back-to-back saturated devices. (b) Oscillator with one device off and the other in triode.*

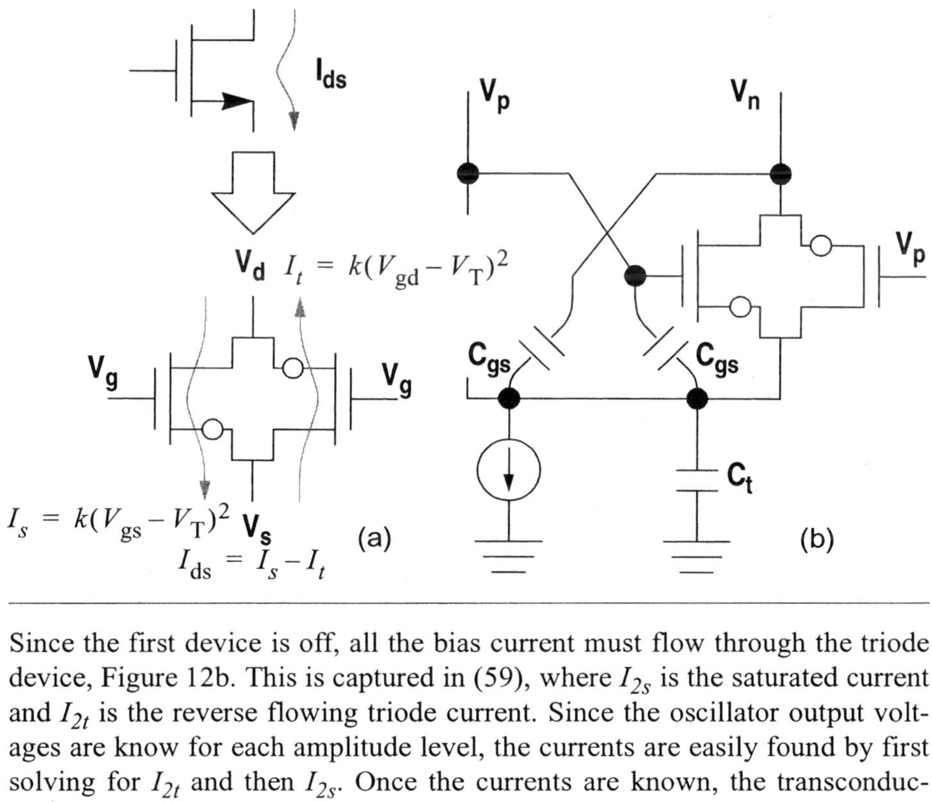

Since the first device is off, all the bias current must flow through the triode device, Figure 12b. This is captured in (59), where I_{2s} is the saturated current and I_{2t} is the reverse flowing triode current. Since the oscillator output voltages are know for each amplitude level, the currents are easily found by first solving for I_{2t} and then I_{2s}. Once the currents are known, the transconductance is calculated.

$$I_0 = I_{2s} - I_{2t} \tag{59}$$

KCL is performed on the small signal model of the circuit and solved in Mathematica for the equivalent impedance. This is given in (60).

$$Z_3 = \frac{2(g_{m2} + s(2C_{gs} + C_t))}{s(4sC_{gs}^2 - (g_{m2} - 2g_{t2})C_t + 2C_{gs}(2g_{t2} + sC_t))} \tag{60}$$

This complicated equation can be simplified for the frequency of interest to (61).

$$Z_3 = \frac{1}{s\left[\frac{2g_{t2}C_{gs}}{g_{m2}} - \frac{C_t(g_{m2} - 2g_{t2})}{2g_{m2}}\right]} \tag{61}$$

The equivalent capacitance is given below.

$$C_3 = \frac{2g_{t2}C_{gs}}{g_{m2}} - \frac{C_t(g_{m2} - 2g_{t2})}{2g_{m2}} \tag{62}$$

Equation (62) is compared to simulation in Figure 13. An ideal current source was used in the simulation to drive the oscillator to unrealistic amplitudes. It is only at these amplitudes that the equivalent capacitance approaches the asymptotic value.

FIGURE 13 *Instantaneous capacitance across amplitude of oscillation for Region 3: C_{gs} = 100 aF, C_t = 100 aF.*

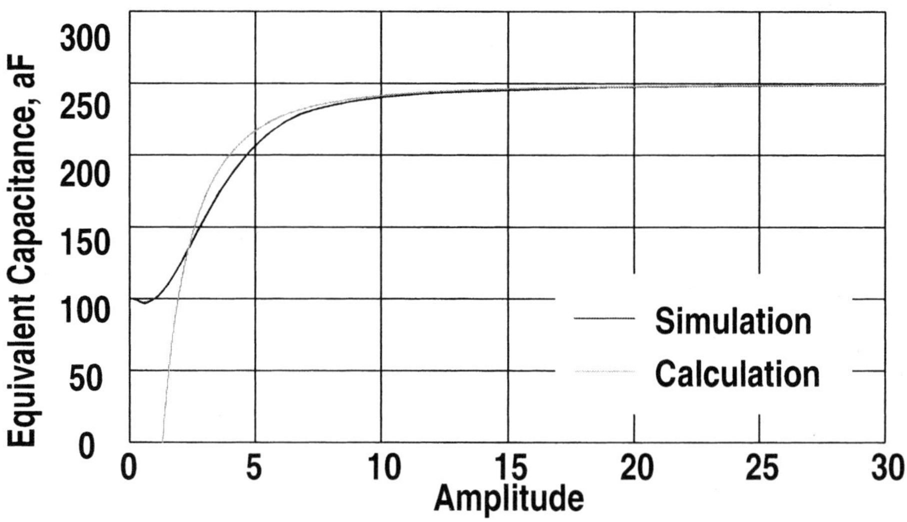

At the peak amplitude, the triode device can be modeled as a short circuit. This is equivalent to setting $g_{m2}=g_{t2}$ in (62) since the drain and the source have the same voltage. Here, (62) simplifies to (63).

$$C_{3,\text{final}} = 2C_{gs} + \frac{C_t}{2} \tag{63}$$

A piece-wise linear model of the instantaneous capacitance, as shown earlier, can be created with the equations for all three regions. However, it was found that fitting a polynomial to the two Region 1 and 2 points gave a better result.

Simple Matlab code generates a sine wave with a given amplitude and maps it to the corresponding instantaneous capacitance using the model just described. The instantaneous capacitance seen by the tank as the circuit oscillates is shown in Figure 14.

FIGURE 14 *Modeled and simulated instantaneous capacitance.*

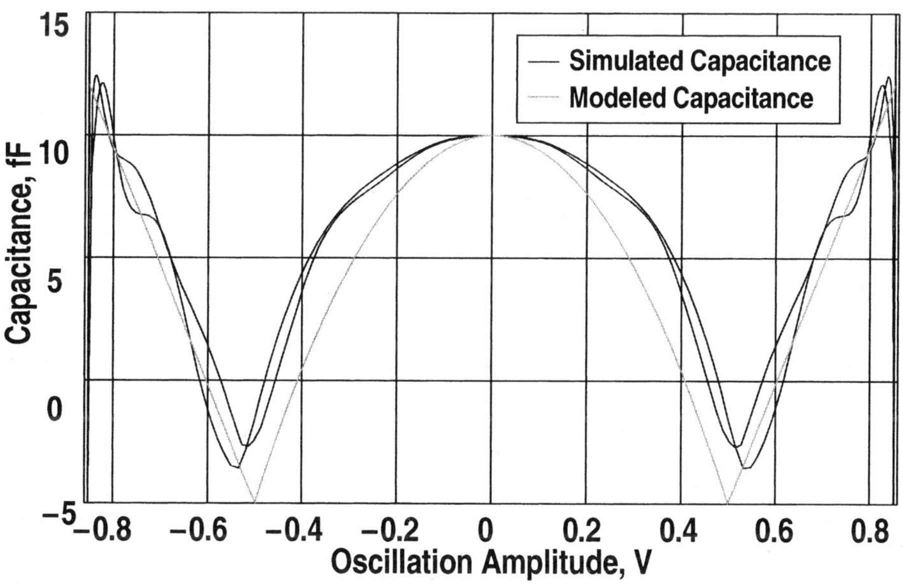

The Fourier Transform of the time domain waveform is taken and the equivalent capacitance is calculated using (54). This is compared to a direct calculation of the equivalent capacitance base on the simulated oscillation frequency as shown in (64).

$$C_{tot} = \frac{1}{L\omega_{osc}^2} \tag{64}$$

$$C_{tot} = C_{tank} + C_{Gr} + C_{mod} \tag{65}$$

The Groszkowski capacitance, C_{Gr}, is calculated by simulating the oscillator without any capacitance and attributing any change in oscillation frequency to this effect. This capacitance is simulated for three separate bias currents. The capacitors are replaced and resimulated for the three bias currents. The additional change in oscillation frequency is attributed to this modulation effect and labeled C_{mod}.

The accuracy of this model is tested for three cases: reasonable tail and gate capacitances, very large tail capacitance, and very large gate capacitance. All three cases are show in table form below.

TABLE 2 $C_t = 10\,fF$, $C_{gs} = 10\,fF$

Bias Current	C_{mod} (Simulation)	C_{mod} (Model)
1390 μA	2.94 fF	6.77 fF
1400 μA	2.96 fF	6.94 fF
1410 μA	2.97 fF	7.10 fF

TABLE 3 $C_t = 100\,fF$, $C_{gs} = 0.1\,fF$

Bias Current	C_{mod} (Simulation)	C_{mod} (Model)
1390 μA	–8.00 fF	–8.56 fF
1400 μA	–7.94 fF	–7.85 fF
1410 μA	–7.88 fF	–7.15 fF

TABLE 4 $C_t = 0.1\,fF,\ C_{gs} = 100\,fF$

Bias Current	C_{mod} (Simulation)	C_{mod}(Model)
1390 μA	36.21 fF	77.79 fF
1400 μA	36.32 fF	78.72 fF
1410 μA	36.44 fF	79.64 fF

While the model seems to generally capture the modulation effect, it fails to do so with sufficient accuracy to accurately predict the equivalent capacitance and thus the modulation coefficient. The model could be refined by using higher order equations to fit the various points. Also, more points across the oscillation curve could be identified to allow a better fit. However, both of these methods make the model less intuitive.

To predict phase noise using this model, first the frequency modulation constant is calculated from the tables above. Using (52) and the amount of flicker noise current present in the bias current leads to the level of phase noise at a given frequency.

In practice, the gate capacitances are time dependent and their values along with the tail capacitance values are not accurately know. RF simulators like SpectreRF can be used to accurately predict the phase noise that occurs in these oscillators.

It is hoped that the model developed in this section adds understanding to the complicated mechanisms for flicker noise up conversion in oscillators.

References

[1] E. Hegazi and A. A. Abidi, "Varactor characteristics, oscillator tuning curves, and AM-FM conversion," *IEEE Journal of Solid-State Circuits*, 2003.

[2] J. Groszkowski, "The interdependence of frequency variation and harmonic content, and the problem of constant-frequency oscillators," *Proceedings of the IRE*, vol. 21, pp. 958-981, 1933.

[3] K. K. Clarke and D. T. Hess, *Communication Circuits: Analysis and Design*. Malabar, Fl: Krieger, 1971.

[4] E. A. Vittoz, M. G. R. Degrauwe, and S. Bitz, "High-performance crystal oscillator circuits: theory and application," *IEEE J. Solid-State Circuits (USA)*, pp. 774-83, 1988.

[5] T. D. Gavra and I. A. Ermolenko, "An ultrashort-wave quartz oscillator with automatic amplitude control," *Telecommunications and Radio Engineering*, vol. 30, pp. 133-134, 1975.

7 Design for Low Flicker Phase Noise

1 Introduction

Despite numerous papers on oscillators, flicker noise has not received much attention. In part, this is because until recently RF oscillators were usually built using the bipolar junction transistors, which have relatively little flicker noise. The need for low flicker noise oscillators became apparent only in CMOS implementations that date to around 10 years ago. Today, almost all integrated LC oscillators are CMOS. This is true even when implementing the RF system in BiCMOS. The fundamental reason is that MOS devices can handle a larger signal swing, which is the most effective means of lowering phase noise.

Flicker noise's behavior in oscillators and mixers is quite distinct from that of thermal noise. Flicker noise is localized in the low frequency portion of the noise spectrum. To affect the phase of a multi-gigahertz oscillator, a frequency translation mechanism must be involved. However, it was shown in an earlier chapter that mixer-based up-conversion of flicker noise can only result in amplitude noise [1]. Flicker noise can only become phase noise through frequency modulation mechanisms. Frequency noise is indistinguishable from phase noise and corrupts communication systems in the same manner. Concepts like noise folding seldom apply to flicker noise in oscillators because, unlike thermal noise, its power is localized in low frequencies.

In a current-biased LC oscillator, the three sources of flicker noise are the current source, the varactor, and the switching pair transistors. Minimization of flicker noise makes use of a variety of techniques discussed in earlier chapters. In this chapter, we will re-visit these techniques from the perspective of minimizing flicker noise.

2 Flicker Noise Minimization

In Chapter 6, the mechanisms by which flicker noise disturbs the oscillator phase were discussed in detail. It is important to recall that while flicker noise of MOS transistors can be lowered by using larger transistors, the amount of frequency noise created by flicker is not a linear function of the transistor's flicker noise. While this might seem a little surprising, it is simply because of the nonlinearity of frequency modulation mechanisms responsible for flicker noise.

Consider the case where only the current source transistor has flicker noise and ignore the nonlinearity of junction capacitances associated with the output transistor. Consider also that the oscillator has no varactor to tune it. In this case, the flicker in the current source transistor does not modulate the frequency of the oscillator output. This is simply because the fluctuation in the bias current only disturbs the amplitude of oscillation, creating amplitude noise. Combined with a nonlinear varactor, the same fluctuation in the source current can disturb the bias point of the varactor, which modulates the frequency of oscillation creating frequency noise as discussed in detail in our varactor treatment in Chapter 8 [2].

Consider a different scenario where the bias current has no flicker noise while the switching pair transistors do. This is a practical case where the bias transistor is replaced with a fixed resistor for example. Assume the common mode point is capacitance-free, including the devices' self-capacitance. In this case, regardless of the amount of flicker noise present in the transistors, the frequency of oscillation is flicker-free. This should be obvious from the discussion in Chapter 6 where the switching FETs flicker noise was shown to modulate the effective capacitance of the active circuit built using the switching pair and the current source. The effective capacitance was shown to depend on the amount of capacitance at the common mode point.

Thermal noise is present in the entire frequency spectrum. Therefore, thermally generated phase noise in an oscillator cannot be eliminated. With clever design, this type of phase noise can be lowered, possibly to its topology-dependent lower bound. Flicker noise on the other hand can be eliminated altogether by intelligently attacking the mechanisms responsible for its up conversion.

In analyzing mechanisms for flicker noise up conversion in Chapter 6, we demonstrated the role the common mode capacitance plays in flicker noise up conversion. The current source device can have an arbitrary *WL* product for any specific *W*/*L* ratio. Larger *WL* leads to lower flicker noise in the transistor itself. However, it comes with larger parasitic capacitance at the common mode point. This increases the flicker noise up conversion gain and the result is more, not less, flicker in the oscillator frequency. It should be remembered that it is always good to increase the channel length of the current source device beyond the minimum feature size. This reduces the excess channel noise coefficient and increases the device's output resistance. However, this reduction in thermal noise is traded for higher up converted flicker noise due to the extra parasitic capacitance at the common mode point.

To alleviate this trade-off, a square-gate layout for the current source is favorable. This layout minimizes the drain junction and maximizes the grounded source junction. This can reduce the capacitive loading on the common-mode point by 50% or more. An array of such devices can be used to implement larger transistors as shown in Figure 1. Simulation shows an optimum point for flicker noise while sweeping the size of the current source transistor. Unfortunately, this optimum point does not coincide with the optimum point for thermal noise.

FIGURE 1 *Layout of the current source device.*

Some researchers argue for inherently low flicker noise topologies using conjectures from the impulse sensitivity function analysis. Based on such analysis, a complementary differential topology can have no flicker noise from the switching devices if the rise and fall times are equal leading to elimination of the even harmonics [3]. However, no analysis has confirmed such conjectures despite the effort of other researchers to using fully symmetric layout [4]. While symmetric layout always helps the suppression of even harmonics in differential circuits, it has not been shown that equal rise and fall times truly lead to suppression of flicker noise. Fully symmetric layout helps by not making flicker noise higher. However, even in simulation with perfectly matched devices, flicker noise shows up in the output spectrum of the VCO.

3 Nulling Flicker Noise

The trade-off between flicker noise of the transistors and flicker noise up conversion gain can be broken with a noise filter [6]. Since the filter inductor splits the common mode point from the current source, the device can be increased in width and length without fear of enhancing the flicker noise up-conversion gain. No special layout is necessary in this case. Proper tuning of the common mode point pushes the flicker noise up-conversion gain from the switching devices into a null. These techniques combined with a switched capacitor arrangement to reduce the oscillator sensitivity leads to large reduction in flicker noise up conversion gain. These techniques combined were implemented by Hoshino et al in a design targeting the FLEX pager specifications [5]. The narrow channel spacing of FLEX requires the very low noise of –110 dBc/Hz at 25 kHz offset. A typical well-designed CMOS oscillator will have an up converted flicker noise corner of around 100-150 kHz. Flicker noise at 25 kHz offset increases the oscillator phase noise by at least 10 log(5) or 7 dB. More than four times the power is required to compensate for the excess noise. Shown in Figure 2 is a sample phase noise measurement with and without the noise filter.

For such stringent specifications, a high-Q off-chip inductor is used in the tank. A combination of a small on-chip inductor and a bond wire inductor is used to implement the noise filter inductor. A large-sized transistor is used for the current source and a 4-bit switched capacitor array is used in parallel with a small varactor. Due to the low supply voltage of 1.5V, the varactor is AC-coupled to the oscillator while biased through a resistor as shown in Figure 3.

FIGURE 2 *Phase noise measurements with and without noise filter.*

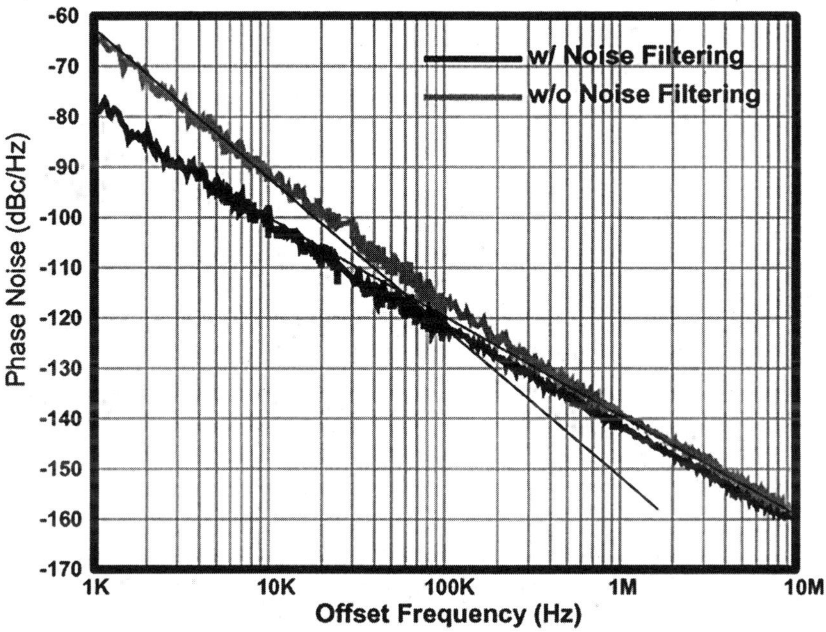

The combination of all the techniques mentioned above resulted in an up-converted flicker noise corner as low as 2 kHz.

However, it is true that the filter nulling effect occurs at a single frequency [5]. Sweeping across the entire frequency range showed an up converted flicker noise corner as high as 60 kHz at the edges of the tuning band. Samples of measured phase noise at various frequencies are shown in Figure 4.

One technique to widen the null in flicker noise up conversion gain was given by Hegazi and Abidi [7]. There, a small-sized switched-capacitor is used in the filter. The auxiliary array is controlled by the same control lines of the main switched-capacitor tank array. This technique allows for multiple nulls by making the noise filter tuning roughly tracking the main tank. This lowers flicker noise for the entire band.

By decoupling the common mode point of the oscillator from the current source, it is possible to increase the size of the current source transistor. Using a channel length that is much larger than the feature size achieves two goals: it

FIGURE 3 *Schematic of a FLEX oscillator.*

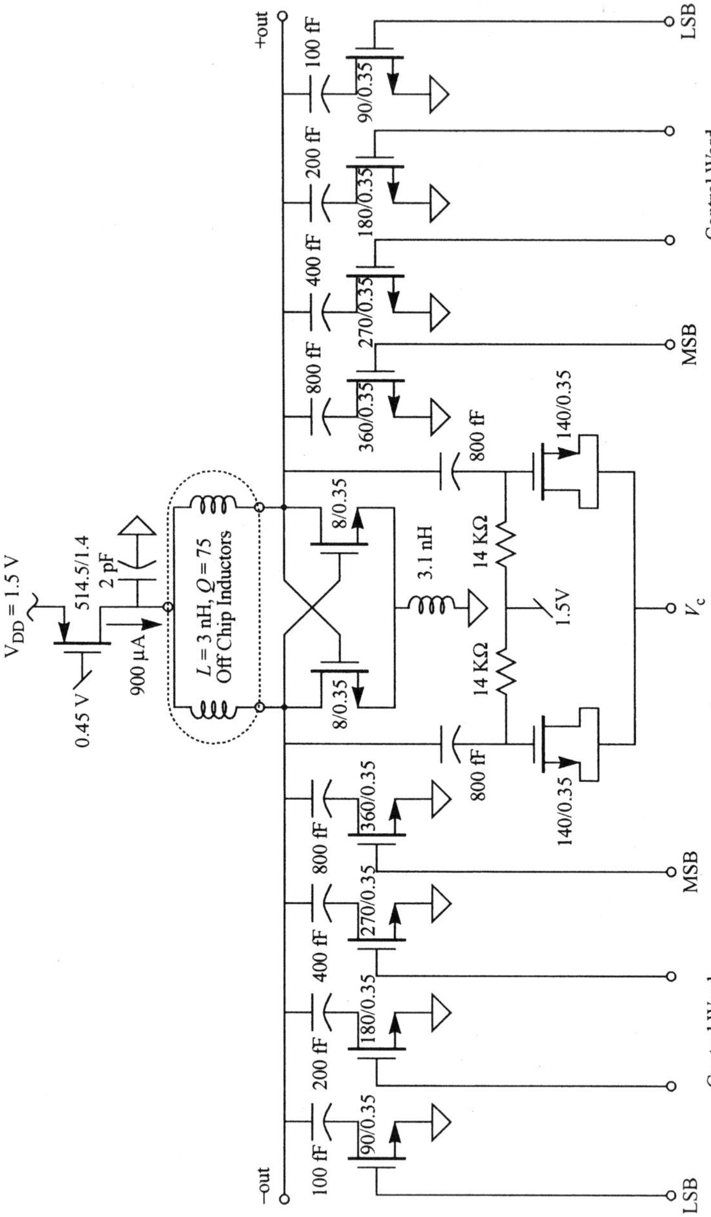

FIGURE 4 *Sample phase noise measurement at the null and at the tuning band limit.*

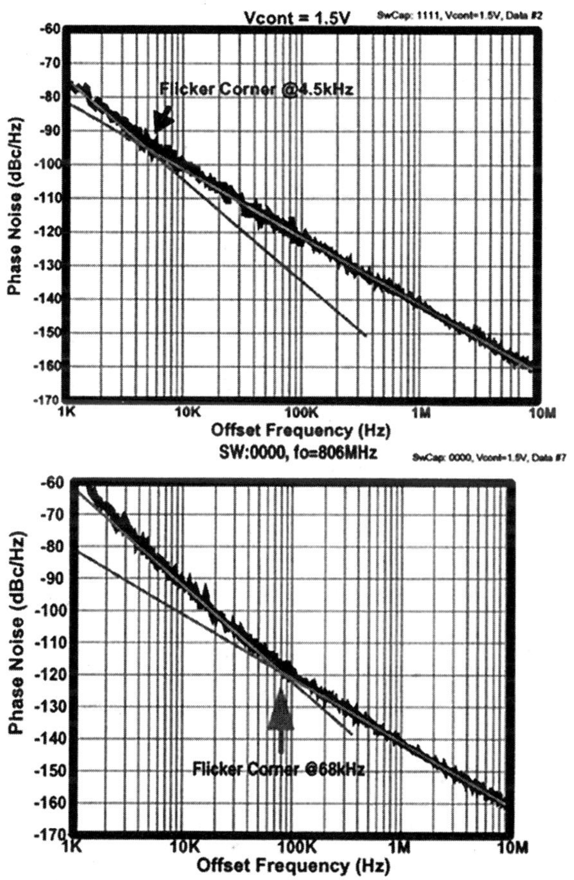

lowers the amount of flicker noise in the transistor, and it lowers the excess channel noise coefficient lowering thermal noise. Finally, as will be presented in Chapter 8, the varactor characteristics can be linearized with the aid of a fixed linear capacitor, either in parallel or in series with the varactor. Such arrangements lower the AM-FM conversion gain [7].

4 Wideband Nulling of Flicker Noise Up Conversion

A technique for nulling the flicker noise up conversion was presented by Ismail and Abidi [8]. This technique, too, relies on eliminating the common mode capacitance. However, they do not tune it out with an inductor. Instead, they realize the elimination of flicker noise through the construction shown in Figure 5. Here, the capacitor C is sized to appear as a low impedance for the fundamental tone that circulates differentially in the cross coupled pair. The capacitor appears as very high impedance to flicker noise in the switching pair. In this topology, there is no common-mode point. With proper choice of the capacitance value, the fundamental oscillation is hardly disturbed and flicker noise is prevented from modulating the oscillation frequency.

FIGURE 5 *Wide-band flicker noise suppression.*

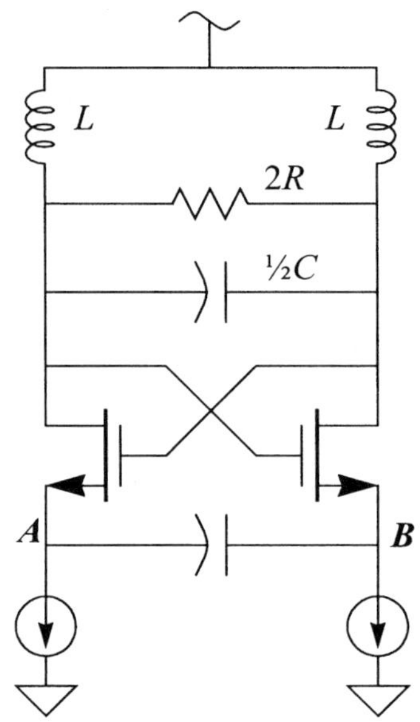

An insightful way to look at flicker noise in the switching pair is by considering it as a slowly varying offset. Without loss of generality, flicker noise can be assigned to one of the two transistors since the switches noise is uncorrelated. First, ignore the parasitics points A and B in Figure 5. The switching transistors are degenerated though a capacitor which reduces the effective negative transconductance of the pair. Therefore, too small a capacitor would prevent start-up of the oscillator. Now how does flicker noise from the switching pair behave? The mixer noise analysis by Darabi proves to be of importance in understanding this circuit. In a regular cross-coupled oscillator, a small flicker noise disturbance of the switching instant modulates the switching moment of the switching pair [9]. The result is duty-cycle modulation of the current fed into the tank. The duty cycle modulation appears at low frequency and at twice the oscillation frequency. In a mixer, this is a mechanism by means of which, flicker noise and clock feed-through make their way to the output (taken at zero intermediate frequency (IF). This was coined the *direct mechanism* [9]. In the context of an oscillator however, this mechanism is of no harmful consequence. Why? Because flicker noise is translated by this mechanism away from the oscillation fundamental. In other words, any flicker noise translated to the tank via this mechanism is rejected by the narrow-band tank characteristic.

In this topology, there is no path for the low-frequency flicker noise to a common mode capacitance. Flicker noise is stored on the floating capacitor such that it re-balances the crossing point and frequency modulation cannot occur. True wideband nulling of flicker noise based on this topology, with the current sources replaced by resistors and a noise filter is added to suppress thermal noise around the second harmonic, was reported by Ismail and Abidi [8].

References

[1] J. J. Rael and A. A. Abidi, "Physical processes of phase noise in differential LC oscillators," in *Proc. of the IEEE Custom Integrated Circuits Conference,* pp. 569-572, Orlando, 2000.

[2] E. Hegazi and A. Abidi, "Varactor characteristics, oscillator tuning curves, and AM-FM noise conversion," *IEEE Journal of Solid-State Circuits,* vol. 38, no. 6, pp. 1033-1039, Jun. 2003.

[3] A. Hajimiri, and T. H. Lee, "Design issues in CMOS differential LC oscillators," *IEEE Journal of Solid-State Circuits,* vol. 34 no. 5, pp. 717-724 May 1999.

[4] B. De Muer, M. Borremans, M. Steyaert and G. Li Puma, "A 2-GHz low-phase-noise integrated LC-VCO set with flicker-noise upconversion minimization," *IEEE J. of Solid-State Circuits,* vol. 35, no. 7, pp. 1034-1038, 2000.

[5] K. Hoshino, E. Hegazi, J. Rael and A. Abidi," A 1.5 V, 1.7 mA, 700 MHz LC CMOS oscillator with no upconverted flicker noise," *Proc. of the European Solid State Circuits Conference*, Villach, 2001.

[6] E. Hegazi, H. Sjoland and A. A. Abidi, "A filtering technique to lower LC Oscillator phase noise," *IEEE Journal of Solid-State Circuits,* Vol. 36 no. 12, pp. 1921-1930 Dec. 2001.

[7] E. Hegazi and A. Abidi, "A 17 mW GSM transmitter and frequency synthesizer in 0.35 mm CMOS," *IEEE Journal of Solid-State Circuits,* Vol. 38 no. 5, pp. 782-792 May 2003.

[8] A. Ismail and A. Abidi, "CMOS differential LC oscillator with suppressed up-converted flicker noise", Proc. of Int. Solid-State Circuits Conference, ISSCC, San Francisco, pp. 98-99, Feb. 2003.

[9] H. Darabi and A.A. Abidi, "Noise in RF-CMOS mixers: a simple physical model," *IEEE Journal of Solid State Circuits,* vol. 35, no. 1, pp. 15-25, Jan. 2000.

8 The Role of the Varactor

1 Fundamentals

Over the oscillation cycle, the varactor capacitance spans a portion of their *C-V* curve that depends on the bias and control voltages as well as the signal amplitude. If the active device nonlinearity is memoryless, the oscillation frequency is determined by the balance of reactive currents in the tank capacitance and inductance [10]. In a differential oscillator, if the parasitic capacitances, other than those across the tank inductance, are negligible, the active negative resistance is nonlinear but memory-free. Furthermore, in an LC oscillator, it is reasonable to assume the oscillation waveform to be quasi-sinusoidal even in the presence of circuit and varactor nonlinearities. This allows the approximation of the frequency of oscillation by

$$\omega_o = \frac{1}{\sqrt{LC_{avg}}}, \tag{1}$$

where L is the effective inductance and C_{avg} is the effective capacitance at balance.

To find the effective capacitance, we resort to basic principles. The oscillator is modeled as a lossy parallel LC tank. The loss is compensated by a negative resistance current that switches from $-I$ to I at the zero crossing of the voltage across the tank. This model captures all nonlinearities of the oscillator while assuming the nonlinear negative resistance is memoryless.

The fundamental component of the negative resistance current flows through the tank loss resistor because the inductor and the capacitor are at resonance at that frequency and so present an open circuit to the fundamental. As such, the amplitude of oscillation is

FIGURE 1 *Parallel resonance tank with switched current drive*

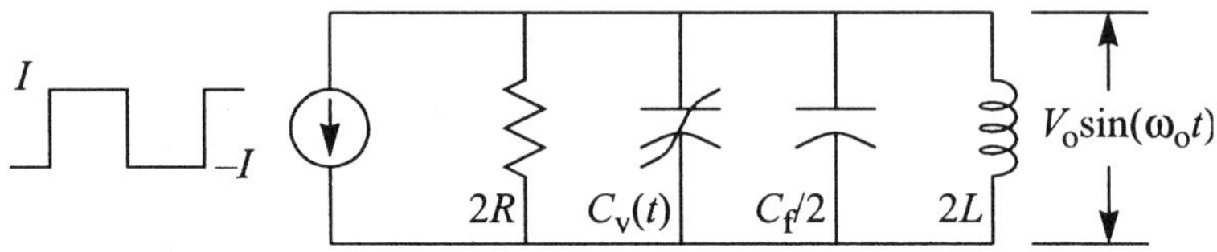

$$V_o = i_o \frac{4}{\pi} Q \omega_o L . \tag{2}$$

Harmonics of the negative resistance current elect to flow in the capacitor rather than the inductor because it provides a lower impedance path. With quasi-sinusoidal operation, harmonics of the negative resistance current cannot flow in the tank resistor which provides a higher impedance path. Note in Figure 2 that current switching affects only the current in the capacitor within a narrow time window around the oscillation voltage-zero-crossings causing more harmonics to flow into the capacitor. For moderate quality factors, 8 or above, the impact of these harmonics on the oscillation frequency can be neglected, in part because the current in the capacitor is Q times the resistor current at resonance and because the switching event occurs in a narrow time window depending on the oscillation amplitude and the switching transistors transconductance.

Having neglected the harmonics of the negative resistance current, the inductor current must balance the capacitor current at all times. Figure 2 shows the inductor and capacitor currents to be equal at all times except at the zero crossing of the oscillation voltage. This moment is when the current is diverted from one side to the other. The sharp transition in current can go through the capacitor but not through the inductor. Under large oscillation amplitude, this transition time is very short and the following applies:

$$i_C + i_L = 0 . \tag{3}$$

The oscillation voltage can be represented in general by its Fourier series expansion as follows

$$V(t) = \sum_n a_n \cos(n\omega t) , \tag{4}$$

FIGURE 2 *Currents in an LC oscillator (simulated).*

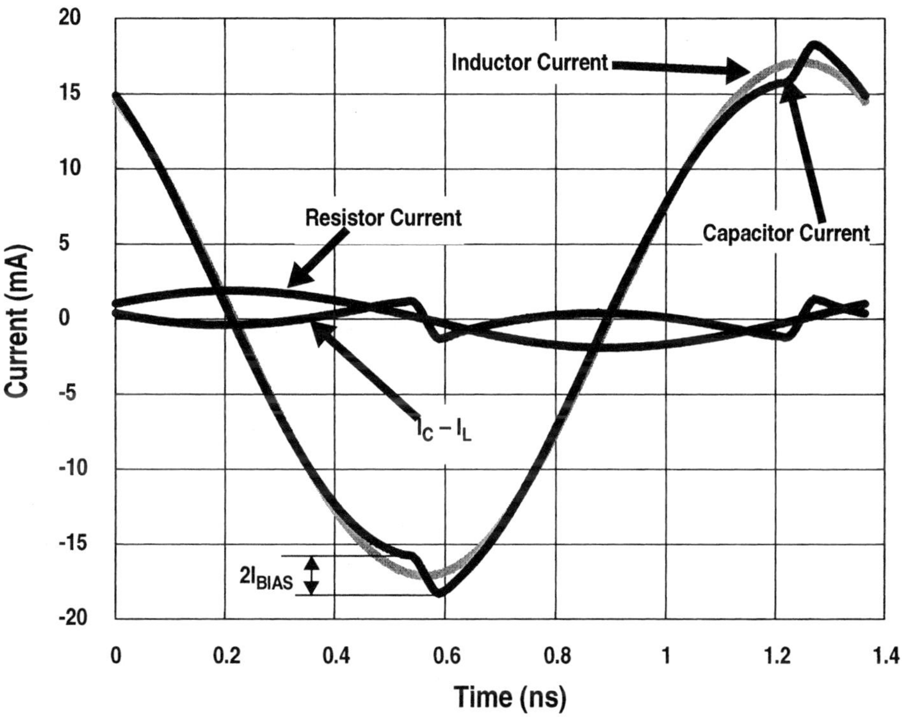

of which only the second term, for $n = 1$, is significant.

The nonlinear capacitance is driven by the sinusoidal voltage and can, in steady state, be represented as a function in time. Note that looking differentially, the capacitance Fourier series becomes a function only of the even harmonics.

$$C(t) = \sum_k C_{2k}\cos(2k\omega t) + \sum_k C_{2k}'\sin(2k\omega t) \qquad (5)$$

We will neglect for the moment all the sine terms and focus only the cosine terms for reasons that will be explained later. The inductor and capacitor currents are given by:

$$i_C = C(t)\frac{d}{dt}V(t)^{\dagger} \tag{6}$$

$$i_L = \frac{1}{L}\int V(t)dt \tag{7}$$

Substituting in the KCL equation

$$\sum_k C_{2k}\cos(2k\omega_o t)\sum_n n\omega_o a_n \sin(n\omega_o t) = \sum_n \frac{a_n}{n\omega_o L}\sin(n\omega_o t) \tag{8}$$

For the fundamental current to balance, the following condition must hold true

$$\omega_o\left(C_0 - \frac{C_2}{2}\right) = \frac{1}{\omega_o L} \tag{9}$$

Where C_0 is the time average capacitance and C_2 is the fundamental of the capacitance waveform, which has a frequency of $2\omega_o$. It is important to note that C_0 and C_2 are frequency independent and need not be evaluated at exactly the oscillation frequency.

$$C_{avg} = C_0 - \frac{C_2}{2} \tag{10}$$

The analysis can be interpreted as follows; fundamental current in the inductor balances with the fundamental current in the capacitor. Differential voltage is odd-symmetric and the capacitance is even symmetric in time. The fundamental of the capacitor current is a sum of products of mixing terms. The time average capacitance mixes with the fundamental of the voltage waveform to produce a current at the fundamental frequency. The second harmonic of the capacitance mixes with fundamental of the voltage derivative to produce a capacitive current at the fundamental frequency and a component at the third harmonic frequency. Owing to the quasi-sinusoidal approximation, (9) neglects the effect of mixing higher harmonics (at $4f_o$, $6f_o$, ...) of the capacitance waveform with higher-order derivatives of the voltage waveform

† An alternate formulation of this equation is considered in Appendix B

(at $3f_o$, $5f_o$, ...). These terms will also contribute a small capacitive current at the fundamental

If the *sine* terms in (5) are considered in addition to the *cosine* terms, they produce a current component that is in phase with the oscillation voltage and in quadrature with both the capacitor and inductor currents, i.e. resistive current, and therefore are not part of the reactive current balance considered in (3). The amount of even harmonics in the differential oscillation voltage is very small and arises fundamentally from circuit mismatches.

This analysis is a reduced form of Volterra series expansion. It is shown in Appendix A that the average capacitance can be evaluated using a different method that, in some cases, is more convenient numerically. The average capacitance is given by:

$$C_{avg} = \frac{A}{\pi\omega_o V_o^2} \tag{11}$$

Where A is the area enclosed by the integral

$$A = \int i dV, \tag{12}$$

and V_o and ω_o are the amplitude and angular frequency of the fundamental component of oscillation. In Section 4 we will look into methods for calculating A both analytically and numerically. Note that the capacitance of the varactor in (6) is the small signal capacitance defined as

$$C = \frac{d}{dV}Q(V) \tag{13}$$

where Q is the charge on the capacitor as a function of the applied voltage V. This definition is equivalent to the small signal capacitance obtained by AC analysis on a SPICE-like simulator.

2 Types of Varactors

The most commonly used varactors today are PN junctions, inversion-mode MOS and accumulation-mode MOS varactors and to a much lesser degree MEMS capacitors. Inversion-mode MOS varactors are most common in

CMOS designs due to their natural compatibility with standard MOS processes as well as their capability of enduring large signal swings. PN junction varactors are more common in bipolar designs with the drawback being their limited signal dynamic range.

All varactors share one common feature; their capacitance varies with the application of a control voltage, generally, in a highly nonlinear manner. As will be shown later, the shape of the nonlinearity has a great impact on the noise performance of the varactor.

3 Varactor Tuning

Harmonic balance mandates equality between the inductive and capacitive currents. The average capacitance is calculated starting from a small signal $C(v)$ expression. Such an expression is often given as $C(v_a,v_k)$ where "a" and "k" subscripts stand for anode and cathode respectively. In this notation the anode is connected to the control voltage while the cathode connects to the oscillator output. The average capacitance is calculated with the aid of (11) as a function of the cathode voltage, which is assumed a sinusoidal waveform. Since the capacitor is a lossless element, once steady state has been reached, the energy stored on it through one-half of the cycle must be extracted from it in another half of the cycle. With sinusoidal stimulus, a linear capacitor responds with a sinusoidal current leading by 90 degrees while a nonlinear capacitor responds with a non-sinusoidal current as shown in Figure 3. Energy is delivered to the capacitor over the first and third quarters of the oscillation cycle while it is extracted throughout the second and fourth quarters. The locus of the voltage versus current in a linear capacitor is a perfect ellipse and in a nonlinear capacitor it is a distorted one depending on the shape of the nonlinearity. For a capacitor driven by a sinusoidal voltage, the increase in voltage amplitude, leads to an increase in current amplitude shown in Figure 4. In a linear capacitor, the area A is proportional to the square of the amplitude such that the capacitance in (11), is independent of the amplitude and so is the frequency of oscillation. In a nonlinear capacitor, the current amplitude is also proportional to the voltage amplitude but the total area is not proportional to the square of the amplitude. Therefore a nonlinear capacitor exhibits a conversion of amplitude variation to frequency variations, Figure 4b.

FIGURE 3 *Energy cycle in a nonlinear capacitor*

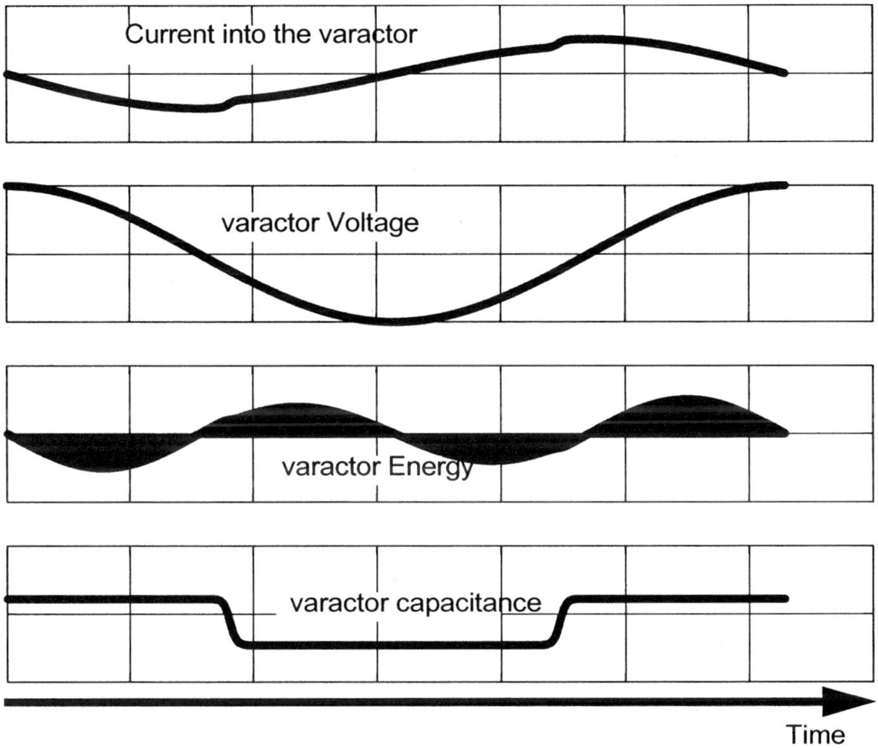

Since the average capacitance is a smooth function, the oscillator-tuning curve is also a smooth function with no abrupt changes in frequency even in an MOS varactor that has a rapid transition in capacitance due to strong inversion.

4 Analytical Evaluation of Noise Sensitivity

The *C-V* curve of a MOS varactor given in [12] defies a simple expression. However, with some approximation, *C-V* characteristics, can be represented by a hyperbolic tangent function that can be curve fitted to the measured or simulated characteristics shown in Figure 5. Moreover, in can be approximated by a step function without much loss of accuracy. The capacitance of

FIGURE 4 *I-V Locus for (a) linear and (b) nonlinear capacitors.*

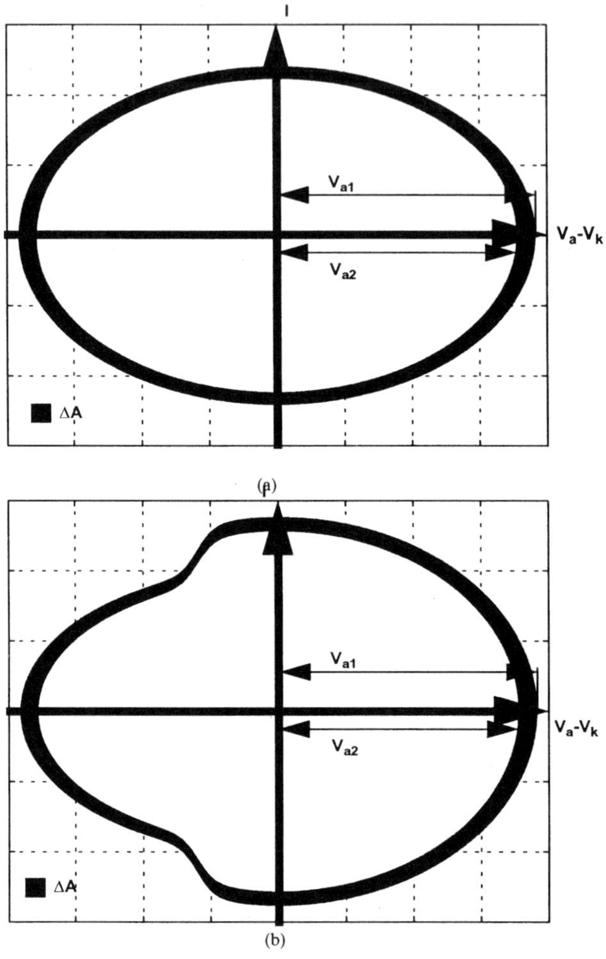

the varactor is therefore switching between the maximum and minimum values at a threshold ε, which is determined by the supply voltage and tuning voltage. With a step-function approximation the capacitance and the sensitivity are now derived.

The small signal capacitance can be approximated by

FIGURE 5 *Varactor's small signal capacitance versus voltage.*

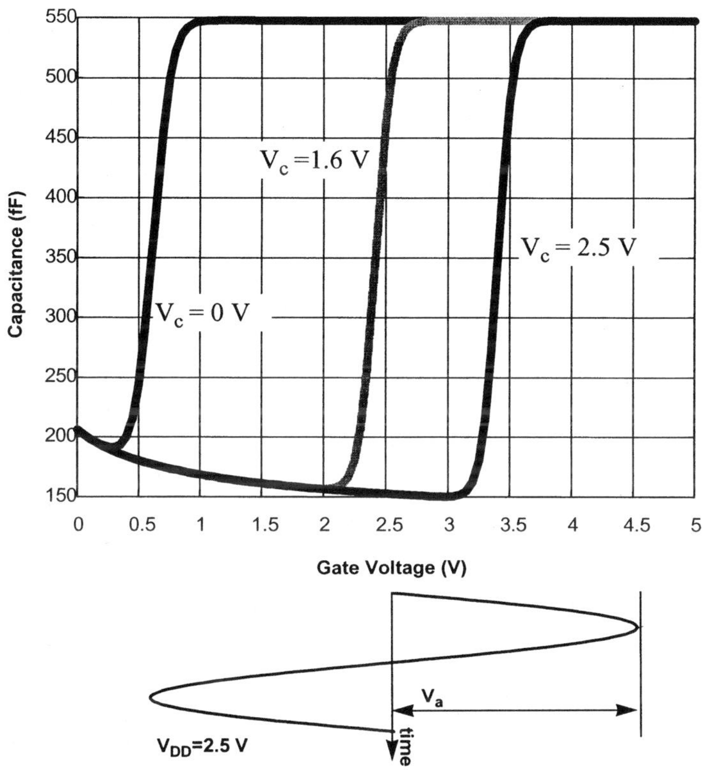

$$C(x_0(t)) \cong \frac{C_{max} + C_{min}}{2} + \frac{C_{max} - C_{min}}{2}(2U(x_0(t)) - 1) \tag{14}$$

where U is the unit step function and x_0 is given by

$$x_0(t) = V_g(t) - V_c(t) + \varepsilon(V_c(t)) \tag{15}$$

where $\varepsilon(V_c)$ is the threshold voltage of the varactor (including bulk effect), is generally close in value to 0.9 V, and is a weak function of the control voltage, V_c. V_g is the gate voltage.

With a step-like capacitance characteristics, the area enclosed by the varactors's *V-I* locus is given by

$$A = 2\left(\int_{-a}^{x_o} idV + \int_{x_o}^{a} idV\right) \tag{16}$$

where

$$\left(\frac{V}{a}\right)^2 + \left(\frac{i}{\omega C_{min} a}\right)^2 = 1 \text{ for } V < x_o \text{, and} \tag{17}$$

$$\left(\frac{V}{a}\right)^2 + \left(\frac{i}{\omega C_{max} a}\right)^2 = 1 \text{ for } V > x_o. \tag{18}$$

where a is the horizontal axis of the ellipse.

The V-I locus therefore, looks like two ellipses merged together at $V = x_o$, Figure 6

$$C_{avg} \cong \frac{1}{2}(C_{max} + C_{min}) + \frac{C_{min} - C_{max}}{\pi}\left(\sin^{-1}\left(\frac{x_o}{a}\right) + \frac{x_o}{a}\sqrt{1 - \left(\frac{x_o}{a}\right)^2}\right) \tag{19}$$

$$k = \frac{\partial C_{avg}}{\partial a} = \frac{C_{max} - C_{min}}{\pi}\frac{2x_o}{a^2}\sqrt{1 - \left(\frac{x_o}{a}\right)^2} \tag{20}$$

All the expressions in (15) through (20) are valid only for $x_o < a$. For $x_o > a$ the expressions saturate to the value at $x_o = a$. Note that k changes from positive to negative as x_o transition from positive to negative, suggesting that it must pass through a zero (a null).

The factor k in (20) represents the tuning sensitivity of the VCO. As shown later, noise fluctuations in the VCO can perturb this average capacitance, and hence the oscillation frequency, resulting in frequency noise sensitivity.

The MOS varactor is characterized by a null in its noise sensitivity curve as shown in Figure 10. This null occurs when the control voltage is such that the varactor remains for exactly half the time at minimum capacitance and the other half at the maximum capacitance. The average capacitance at this particular tuning voltage is constant regardless of the amplitude of oscillation. Therefore, assuming symmetric C-V curve around an inflection point, the var-

FIGURE 6 *Voltage-current loci for various control voltages.*

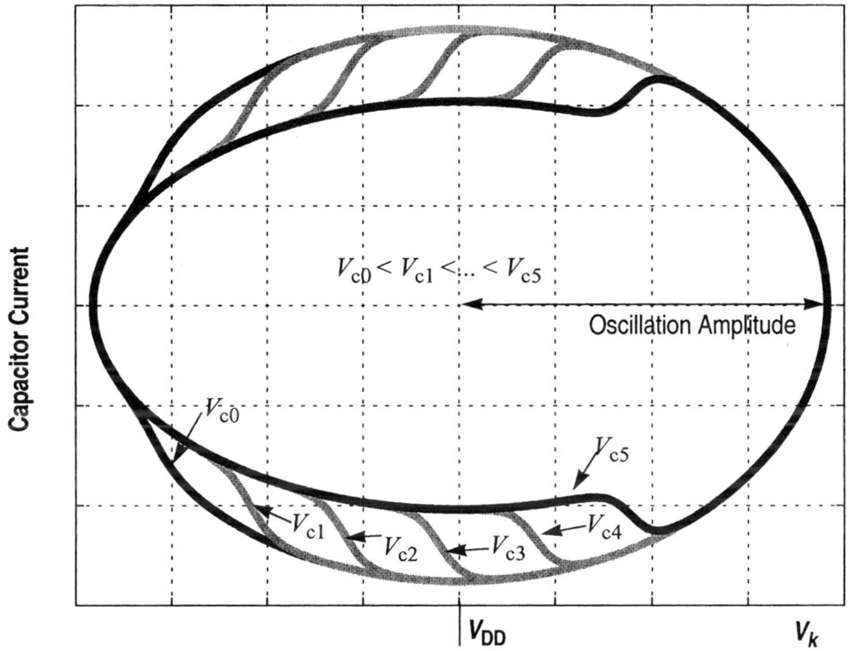

actor shows zero sensitivity to amplitude variations and average capacitance is the numerical average between the maximum and minimum values. The condition for the null is given by:

$$V_c + \varepsilon(V_c) = V_{go}, \tag{21}$$

or simply:

$$x_o = 0 \tag{22}$$

where V_{go} is the common mode level of the gate voltage, equal to V_{DD} in the tail-biased topology.

For PN junction diode varactors, the $C(v)$ curve is smoother than that of MOS varactors, leading to less sensitivity to amplitude variation. However, the analytical evaluation of the integration in (16) is far more complicated and must be evaluated numerically. Since the capacitance curve does not have an inflection point, there exists no null in the sensitivity curve.

5 AM-to-FM Noise Conversion

In a current-biased oscillator, low frequency noise from the current source is sampled by the commutating switching pair. This is up-converted as two correlated sidebands around the fundamental creating pure amplitude noise [1][2][11].

The use of a noise filter does not prevent the AM modulation as it allows low frequency noise from the current source to pass almost unattenuated [3]. The nonlinear varactor capacitance is modulated by the noise-induced amplitude variation leading to change in frequency of oscillation. The spectra of FM noise and phase noise are indistinguishable on a spectrum analyzer and degrade communication systems in the same manner. In this section we develop a simple model of the AM-to-FM conversion process.

For an AM noise component to up-convert around the fundamental frequency it needs to have originated at low frequency. This is a slowly varying process, therefore, only the average behavior of the varactor, namely its average capacitance, determines the AM-to-FM noise conversion gain. AM noise of interest can therefore be regarded as low-frequency envelope variations that allow the oscillator enough time to change its frequency before the noise level changes. Finally the average capacitance is assumed a linear function of the noise voltage at the tank terminals. The average capacitance is however, a nonlinear function of the output signal voltage at the tank. This is the typically adopted small signal assumption in treating noise and can be expressed as follows:

$$C(t) = C_{\mathrm{f}} + C_{\mathrm{avg}} + kv_{\mathrm{n}}(t) \tag{23}$$

Where C_{f} is the fixed capacitance at the oscillator output node, C_{avg} is the average varactor capacitance and k is the sensitivity of C_{avg} to amplitude variations given by the partial derivative of the average capacitance with respect to amplitude.

With the aid of (1) and given that the noise voltage is small, it can be easily shown that the oscillation frequency can be approximated by

$$\omega_{osc} \cong \omega_o\left(1 - \frac{kv_n(t)}{2C_T}\right) \tag{24}$$

where C_T is the sum of the fixed and varactor average capacitances and ω_o is the nominal oscillation frequency.

With the aid of narrowband FM, NBFM, theory, phase noise due to capacitance modulation is given by

$$\mathcal{L}(\omega) = J_1^2\left(\frac{\omega_o k}{\omega}\frac{v_n}{4C_T}\right) \cong \left(\frac{\omega_o}{\omega}\right)^2\left(\frac{k}{4C_T}\right)^2\overline{v_n^2} \tag{25}$$

where J_1 is first order Bessel's function of the first kind.

The current source device, is accountable for most of the AM noise in a VCO [2]. The switching devices produce pure phase noise. Therefore, the current source device is responsible for most of the AM-to-FM noise.

The noise voltage due to a noise current i_n in the current source is given by the following expression

$$v_n = i_n\frac{2}{\pi}Q\omega_o L \tag{26}$$

Note that the noise current represents the sum of thermal and flicker noise currents as they are referred to the output of the current source transistor.

$$\overline{i_n^2} = 4kT\gamma g_{m_{bias}} + \frac{K_f}{f}\frac{g_{m_{bias}}^2}{WLC_{ox}} \tag{27}$$

where K_f is the flicker noise coefficient. The remaining part of the analysis presented is the calculation of the sensitivity parameter, k. It can be calculated by one of two methods, analytically or numerically as shown earlier.

6 Tuning and Supply Sensitivity

The tuning sensitivity of the varactor can be evaluated using the average capacitance expression in (19). The result is

$$k_t = \frac{\partial C_{avg}}{\partial V_c} = -\frac{\partial C_{avg}}{\partial x_o}, \tag{28}$$

$$k_t = \frac{(C_{min} - C_{max})}{\pi}\frac{2}{a}\sqrt{1 - \left(\frac{x_o}{a}\right)^2}, \tag{29}$$

which is a monotonic function of x_o, unlike the noise sensitivity, which changes signs at $x_o = 0$.

Now the tuning sensitivity of the VCO can be predicted accurately as follows:

$$K_{VCO} = -\frac{\omega_o C_{avg}}{2}\frac{\partial C_{avg}}{\partial V_c} \tag{30}$$

Power supply noise is also a source of phase noise in the oscillator. Noise on the supply line around the even harmonics of oscillation is responsible for phase noise at the output. In a previous work, [4] it was shown how the noise filter can remove the noise around the second harmonic. In the presence of varactor nonlinearity, low frequency fluctuations on the supply line modulate the varactor capacitance and result in FM noise skirts. The noise filter reduces the translation gain of low frequency noise to phase noise but cannot stop this noise from modulating the varactor. To quantify this effect, we note that the average capacitance of the varactor is a function of the voltage difference between its terminals. In a tail-biased VCO for example, the varactor is connected from one side to the supply. Therefore, the power supply noise sensitivity is opposite of the tuning sensitivity,

$$k_{supply} = \frac{(C_{max} - C_{min})}{\pi}\frac{2}{a}\sqrt{1 - \left(\frac{x_o}{a}\right)^2}. \tag{31}$$

Therefore, phase noise due to low frequency supply fluctuations is given by

$$\mathcal{L}(\omega) = \left(\frac{\omega_o}{\omega}\right)^2\left(\frac{k_{supply}}{4C_T}\right)^2\overline{v_{n_{supply}}^2}. \tag{32}$$

To account for low frequency noise on the control line, the RMS noise power in a 1 Hz bandwidth in this equation should be replaced by the sum of the noise power spectral density on the control line and on the supply line since they both have the same gain to the output.

The crossing point corresponds to the null in the noise sensitivity plot as shown in the 3-D plot shown in Figure 7. This is because, at this tuning voltage, the change in the amplitude of oscillation does not yield a change in the average capacitance as described earlier. Note that the higher the amplitude, the lower the noise and tuning sensitivities become. Larger oscillation amplitudes lead to more averaging of the nonlinear capacitor and hence lower tuning sensitivity as shown in Figure 8. However, the tuning sensitivity variation is also reduced which means that the oscillator tuning sensitivity remains almost constant across the tuning range, resulting in a more linear frequency-tuning curve. In addition, larger bias current brings the oscillator closer to amplitude limiting and AM noise is less effective in perturbing the oscillation frequency. As a result, the noise sensitivity coefficient is reduced as shown in Figure 7.

FIGURE 7 *AM-FM noise sensitivity.*

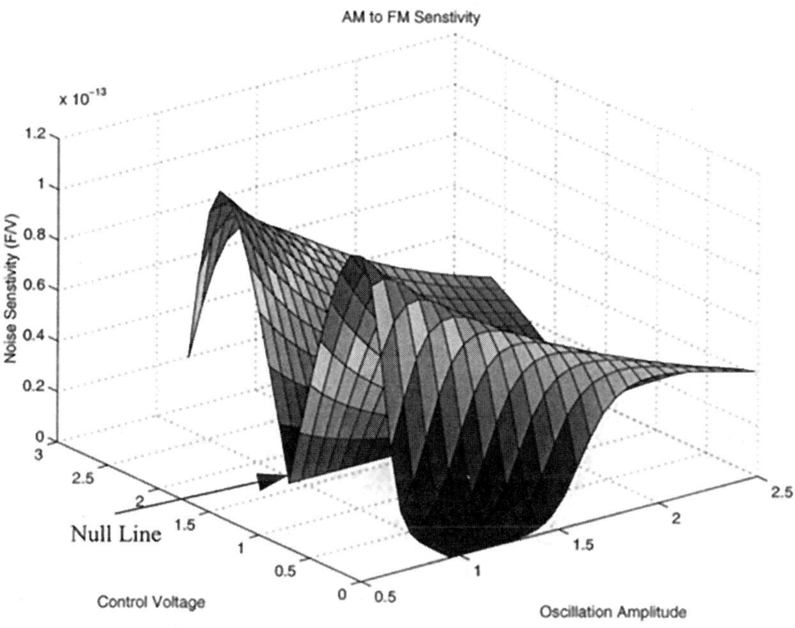

7 Measurements and Simulation Results

The analytical results and physical insights developed so far are validated in two different ways: by direct measurement on fabricated integrated circuit oscillators and by SpectreRF simulations. The first set of measurements is carried out on varactors in a BiCMOS6G 0.35 μm process from STMicroelec-

FIGURE 8 *Tuning sensitivity.*

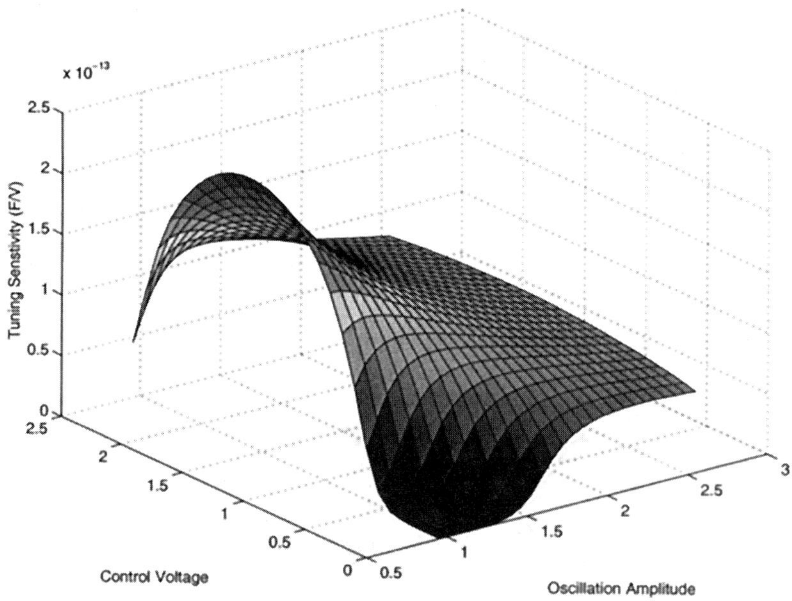

tronics. NMOS inversion mode varactors are used in an LC oscillator along with switched capacitors for discrete tuning. The varactor consists of 64 parallel NMOS transistors of 5μ/0.35μ each. A three bit switched capacitor array of a 60 pF unit capacitor implements the switched-tuning network. The total fixed capacitance is 1060 fF in parallel with a 13 nH inductor. The supply voltage of this tail-biased VCO is 2.5 V. We compare the measured frequency-tuning curve with analytical predictions. The oscillator is operated in the current limited regime, where the amplitude is proportional to tail current.

A tail current less than 4 mA guarantees that the supply voltage does not limit the amplitude. To verify the dependence of the varactor's effective capacitance on the oscillation amplitude, the frequency-tuning curve is measured at two different amplitudes set by the tail currents. As expected, the two curves are different (Figure 9a). The tuning sensitivity lowers with amplitude as the oscillation sweeps across a greater portion of the varactor characteristic, averaging it more. This test oscillator has a noise filter. In the presence of a noise filter, the effects of higher harmonics described in [10] are minimized [3].

FIGURE 9 *Measured VCO tuning curves versus model.*

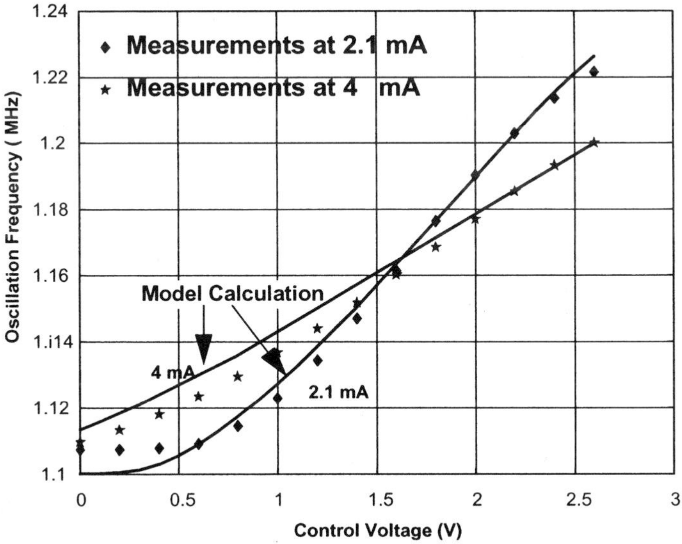

We predict the frequency-tuning curve with the formulas given above by approximating the MOS varactor characteristic with a step. In spite of this simplification, the prediction lies close to the measurement. The tuning sensitivity (the slope) predicted by the model is notably accurate. Next, we search for the predicted nulls in AM-FM conversion. Low-frequency noise in the oscillator's tail current first up-converts into AM sidebands [2], and the varactor converts AM into FM. We study this effect by simulating the periodic transfer function with SpectreRF from low frequencies (1 kHz) in the tail current to the phase noise sidebands in the oscillator output (Figure 10b). We compare the simulated transfer function versus the gate overdrive voltage, V_{eff} at I_{bias}= 2.3 mA with analytical predictions. Although simulation does not yield perfect nulls, there is a clear drop in simulated sensitivity at the two anticipated bias points. Note that the translation gain in SpectreRF is scaled by the signal power and then compared to that predicted by (25). The translation gain null at low control voltage is dependent on signal amplitude (current and tank quality) while the null at 1.6 V is not. The null at 1.6 V depends only on the common-mode voltage on the varactor gate as well as the threshold voltage. Note that the discrepancy between SpectreRF simulation results and those of the model are higher at low values of V_c. This is mainly because the hand calculations assume $\varepsilon(V_c)$ to be always 0.9 V while at low values of V_c, bulk effect is less pronounced and $\varepsilon(V_c)$ is smaller.

The nulls are also verified by measurements. It is difficult to directly measure the small signal response simulated in SpectreRF, but we can deduce the presence of nulls by superimposing the frequency-tuning characteristics at the three different tail currents, I_{bias}= 2.1, 3, and 4 mA (Figure 10a). Over this range of currents, the amplitude is proportional to I_{bias}. Measurements show that all three tuning curves intersect at V_c of 0 and 1.6 V. This means that at both intersection points the *frequency* is independent of *amplitude*, that is, AM does not cause FM. The tuning voltage at each intersection point is as predicted, which confirms the analysis.

Finally, we validate the phase noise in an oscillator caused by AM-FM conversion in the varactor. This is a difficult experiment because in a practical circuit many mechanisms are at work simultaneously to produce phase noise. We have shown that in a differential oscillator, FM through the varactor is only one of three different mechanisms responsible for the up-conversion of low frequency ($1/f$) noise around the carrier [2]. Therefore, for our purposes, the required test oscillator is one with a strong varactor (large K_{AMFM}) to emphasize AM-FM conversion, while at the same time is designed to suppress the two other mechanisms responsible for up-converting low frequency noise. Close-in phase noise will also appear through additive mechanisms, but in a differential oscillator this arises from device noise near the oscillation frequency and its multiple as we showed earlier in the discussion on thermally-generated phase noise. Since this additive noise is originated at high frequencies (multiples of the oscillation frequency), it must be thermal not flicker noise. In CMOS, noise originating at low frequencies is immediately discerned by the fact that the slope of its spectrum is 10 dB/decade higher than that for white noise. This makes it easy to discriminate between phase noise due to additive mechanisms and due to AM-FM conversion; the slope of the former is 20 dB/decade, whereas it is 30 dB/decade for the latter.

The test oscillator is fabricated in 0.35μm CMOS with STMicroelectronics' BiCMOS6M process. To suppress up-converted flicker noise originating in the differential pair, the parasitic capacitance at the common mode point is lowered by laying out the current source FET as a group of annular gates; this lowers the capacitance of the drain junction, which lies inside the annulus. The differential pair FETs are themselves of small size, 15 μm/0.35 μm. While this raises the input-referred flicker noise of the FETs, their smaller capacitance more than compensates by lowering the up-conversion gain [2].

Simulations confirm that the dominant mechanism for flicker noise up-conversion in this VCO is varactor nonlinearity. The tank inductance is 17 nH with a Q of 8. The oscillation frequency ranges from 1040-1140 MHz over the full tuning range from 0 to 2.5 V. Using the formulas given by Rael and Abidi [2], phase noise at 3 MHz offset that arises from white noise only is predicted to be –142 dBc/Hz; this is verified by direct measurement. By straight forward extrapolation, we can say that at 10 kHz offset white noise accounts for a phase noise of –91 dBc/Hz. A higher phase noise than this must then be due to up-converted $1/f$ noise. Figure 11 shows the measured phase noise at 10 kHz offset across the full tuning range versus the prediction of (18). At a tuning voltage of $V_{DD} - V_T$, which is 1.6 V in this circuit, the measured phase noise falls to the floor imposed by white noise. The analytical prediction is based on flicker noise in the bias current in (19) and the well-known expression for the spectral density of MOS flicker noise

$$\overline{i_n^2} = \frac{K_f}{f} \frac{g_m^2}{WLC_{ox}} \tag{33}$$

where we use $K_f = 1.58 \times 10^{-24}$ W. Model accuracy is improved by including the variations of V_T due to changing V_c. The control voltage is taken from a battery, which has very low noise; thus the main source of noise is the tail current. This mechanism of phase noise can vary by up to 10 dB at low offset frequencies if the varactor is nonlinear; that is, its C-V curve departs from a straight line. Linearity may be improved by connecting a linear capacitor either in series or in parallel with the varactor. Series connection achieves linearization at the expense of tuning range. With the parallel connection the loss in tuning range can be regained using a mixed signal control over the VCO.

8 Discussion

AM-to-FM noise conversion is an inevitable consequence of the nonlinear varactor characteristics. Different varactors show different AM-to-FM noise conversion coefficients. PN junction varactors have the smoothest *C-V* curve amongst the three most common varactors and therefore generate the least amount of AM-to-FM conversion. Junction capacitors of the switching transistors as well as any other nonlinear capacitances across the tank convert AM noise into FM noise. However, the dominant nonlinear capacitance is typically that of the tuning varactor. The noise conversion can be mitigated through linearization of the varactor characteristic. This can be done using switched capacitor arrays for discrete tuning to reduce the contribution of the nonlinear capacitor to the total tank capacitance [13]. Another alternative is to add a linear capacitor in series with the varactor. The parallel solution requires mixed signal control of the VCO whereas the series solution comes at the expense of tunability. It is worth mentioning that adding a series linear capacitor is typical for PN junction varactors to lower the signal swing seen by the PN junction and avoid forward bias. One other effective method is noise filtering in LC oscillators [3]. It is important to recognize that the presence of the noise filter allows the use of extremely large current source devices that have a channel length much larger than the technology's feature size. The large size current source has less flicker noise and therefore creates less AM noise in the first place. Noise filtering has proven useful for the total elimination of flicker noise in LC oscillators [4].

Appendix A

The basic premise behind finding an average capacitance is to use small signal capacitance simulations, which are performed on a driven capacitor to approximate the behavior of the capacitor in an autonomous oscillator. Therefore, the characterization of the capacitor to obtain the average capacitance should be frequency independent. The way to do that is to assume quasi-sinu-

FIGURE 10 *Measured VCO tuning curves meet at nulls of the AM-FM conversion (top). PXF gain at 1 kHz from the tail current source to phase noise (bottom).*

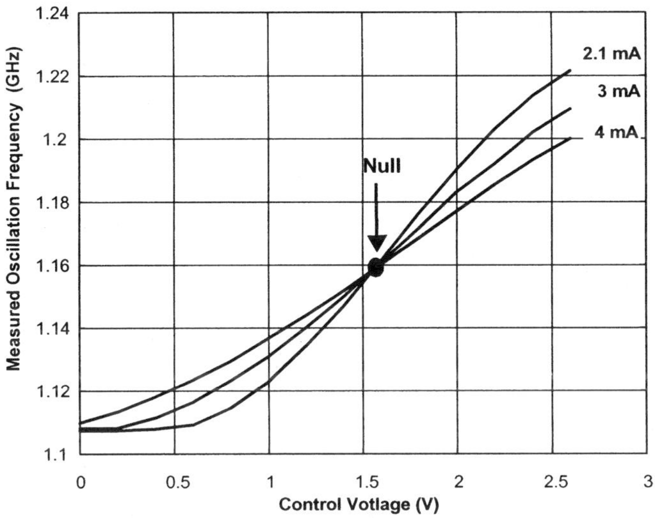

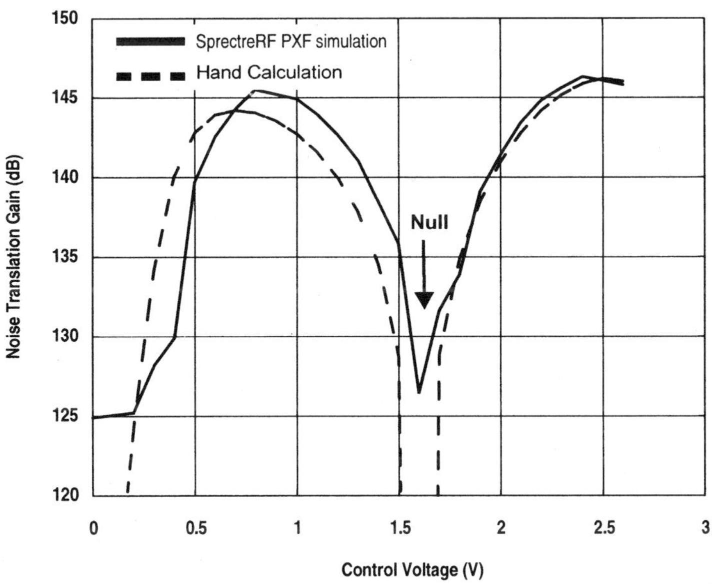

FIGURE 11 *Measured and analytical prediction of phase noise at 10 kHz offset.*

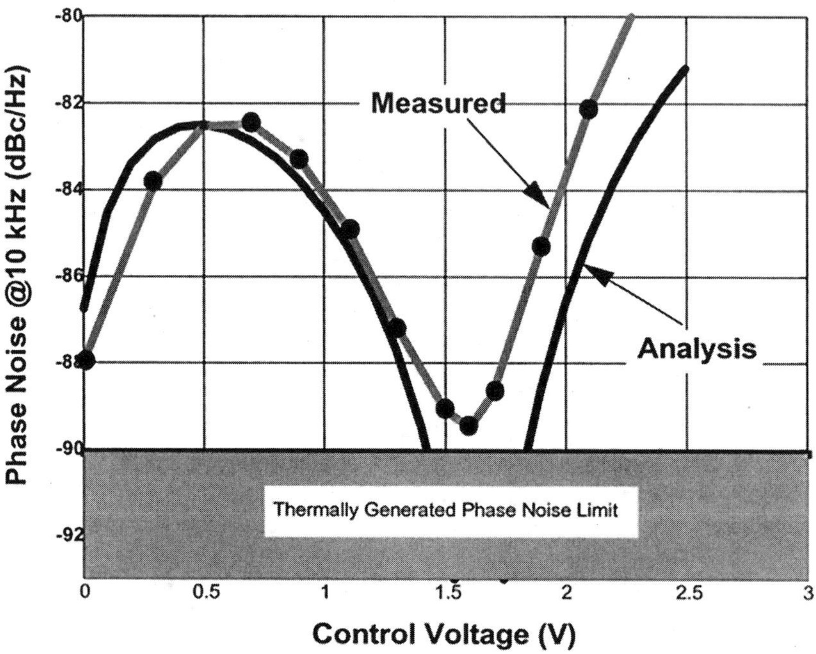

soidal operation of the oscillator. What this really means is that we assume most of the nonlinearity of the circuit comes from the varactor not from the negative resistance. This is not so true for low quality oscillators. However, when the oscillator is of reasonable quality, which is currently feasible in fully integrated oscillators, this approximation holds. The integral

$$\oint i dV = \oint C(V)\frac{dV}{dt}dV = \int_0^T C(V)\left(\frac{dV}{dt}\right)^2 dt \tag{34}$$

can be proven proportional to the average capacitance calculated in Section 1. With the quasi-sinusoidal assumption, the voltage V can be expressed as

$$V(t) = a_1\cos(\omega t) \tag{35}$$

Then

$$\oint i dV = A = \int_0^T C(t)\omega_o^2 a_1^2 (\sin(\omega t))^2 dt \tag{36}$$

$$A = \frac{\omega_o^2 a_1^2}{2}\left[\int_0^T C(t)dt - \int_0^T C(t)\cos(2\omega t)dt\right] \tag{37}$$

These integrals are the exact definitions of the coefficients of the Fourier series of the capacitance,

$$A = \frac{\omega_o^2 a_1^2}{2} T\left(C_o - \frac{C_2}{2}\right). \tag{38}$$

From which is follows that the average capacitance is given by

$$C_{avg} = \frac{A}{\pi\omega_o a_1^2} \tag{39}$$

Appendix B

In analyzing large signal behavior of nonlinear capacitors, the question always arises about the definition of capacitance.

The definition of capacitance is

$$C \equiv \frac{dQ}{dV}. \tag{40}$$

First consider a nonlinear capacitor subjected to a large single tone voltage. This is not exactly the case of a varactor in an LC oscillator because of the presence of noise. For the moment, we shall neglect noise and consider only the large oscillation signal.

The charge on the capacitor is described in general by a nonlinear function:

$$Q = f(V) \tag{41}$$

Using the first three terms of the Taylor series expansion of *f*, we can approximate Q with

$$Q = a_o + a_1 V + a_2 V^2 . \tag{42}$$

Assuming a single tone sinusoidal voltage, V is given by

$$V = A\sin(\omega t) . \tag{43}$$

The varactor current is given as:

$$i = \frac{dQ}{dt} = \omega_o a_1 A\cos(\omega_o t) + 2\omega_o a_2 A^2 \cos(\omega_o t)\sin(\omega_o t) \tag{44}$$

Now evaluate the following expression:

$$\frac{\partial Q dV}{\partial V dt} = \omega_o a_1 A\cos(\omega_o t) + 2\omega_o a_2 A^2 \cos(\omega_o t)\sin(\omega_o t) \tag{45}$$

This is identical to (44). This result can be shown to apply to any order of the Taylor series expansion of *f*. This conforms to the capacitance definition adopted in (6). The former analysis shows that the instantaneous capacitance of an arbitrarily nonlinear capacitor driven by a noise-free single tone is given by:

$$C(t) = \frac{\partial Q}{\partial V} \tag{46}$$

Next consider small signal perturbation is added to the varactor voltage V. We use upper-case V for the large signal and lower-case v for the perturbation. The small v can model a noise disturbance to the periodic steady state varactor voltage.

Again, the charge on the capacitor is given by:

$$Q(t) + q(t) = f(V + v) \tag{47}$$

Using Taylor approximation leads to:

$$Q(t) + q(t) = f(V) + v\frac{df}{dV} \tag{48}$$

Using (41) leads to

$$q(t) = v(t)C(t) \tag{49}$$

where

$$C(t) = \frac{dQ}{dV} \tag{50}$$

and

$$i_C = \frac{d}{dt}q(t) = C(t)\frac{dv(t)}{dt} + v(t)\frac{dC(t)}{dt} \tag{51}$$

Which is different from (6).

The small noise perturbation voltage in (51) produces a perturbation charge that is linearly proportional to v. This is what can be expected from a small signal approximation around a periodic steady state solution.

Comparing (51) to (6), it turns out that our previous analysis missed the term $v(t)\, dC(t)/dt$. This term may potentially up-convert some low-frequency noise and alter the results we obtained earlier. Note that this extra term appears only in the presence of a small perturbation signal $v(t)$ added to the large signal $V(t)$.

The question is whether this is the case under investigation for AM-FM noise conversion in varactors. To answer this question, take a closer look at (51). The small signal perturbation $v(t)$ represents flicker noise and hence it has energy at low frequencies only. This is also the case for its derivative $\mathrm{d}v(t)/\mathrm{d}t$. The terms $C(t)$ and $\mathrm{d}C(t)/\mathrm{d}t$ have energy around $2\omega_0$, $4\omega_0$, ..., $2n\omega_0$. The mixing of these terms in (51) cannot result in noise around the fundamental. To the contrary, the model in (6) is based on the mixing of $C(t)$ with $V(t)$ which is the large signal oscillation voltage. $V(t)$ has energy around ω_0, $3\omega_0$, ..., $(2n-1)\omega_0$. Therefore, $V(t)$ is modulated in amplitude by flicker noise, the average capacitance can be modulated as shown in (23) and (24).

The effective capacitance of the varactor is calculated with only the VCO large signal applied to the varactor. In presence of flicker noise the situation will not change. Flicker noise is very slow compared to the oscillation frequency and the oscillator has an ample time to stabilize its amplitude before a new sample of flicker noise is applied. Note that the model used to calculate frequency fluctuations relies entirely of the concept of average capacitance perturbation which is only valid if the perturbation is much slower than the oscillator dynamics.

Appendix C

In the analysis of circuits involving nonlinear capacitors, there is always confusion about whether to use the capacitance definition in (46) or the following definition:

$$C = \frac{Q}{V} \tag{52}$$

The reader is cautioned here that V in the above expression is the large signal voltage not the noise perturbation. Therefore (52) is not the same as (49). It follows that the capacitive current is given by:

$$i_C = \frac{d}{dt}Q(t) = C(t)\frac{dV(t)}{dt} + V(t)\frac{dC(t)}{dt} \tag{53}$$

We now express the capacitance $C(t)$ in terms of its Fourier series harmonic content

$$C(t) = \sum_k D_{2k}\cos(2k\omega t) + \sum_k D_{2k}'\sin(2k\omega t) \tag{54}$$

Note that we use here the symbol D instead of C to describe the harmonics so that we don't confuse them with those used in the original analysis (see (8)).

The reader can carry the analysis as before by substituting from (54) above into (53) and equate the fundamental of the capacitive current with that of the inductive current. The resulting expression for the effective capacitance is given as follows:†

$$C_{eff} = D_o + D_2/2 \tag{55}$$

Comparing (55) with (10), the two expressions for the effective capacitance seem very similar except for the plus (+) sign instead of the minus (–) sign.

† The expression in (55) was derived by Z. Tang who substituted (54) into (53) but ended with the wrong expression $C_{eff} = C_o + C_2/2$ instead of (55). [14]

We now show that $D_2 = -C_2$ hence prove the two approaches identical.

Let's start by the area enclosed by the *i-v* locus of the capacitor which is given by:

$$\oint i dV = \oint \frac{dQ}{dt} dV = \oint \frac{d}{dt}((c(t)V(t)))dV \qquad (56)$$

Again if we make the substitution from (54) and carry out the analysis as before, it can be shown that the area is given by:

$$A = \frac{\omega_o^2 a_1^2}{2} T\left(D_o + \frac{D_2}{2}\right) \qquad (57)$$

where A is the area enclosed by the *i-v* locus

Compare (57) above with . The two right-hand sides have to be identical. It follows that

$$D_o + D_2/2 = C_o - C_2/2 \qquad (58)$$

Hence, the two analyses are identical. Moreover, the expression for the effective capacitance is identical using either definitions of the capacitance.

Finally, we answer the following question: why choose the capacitance definition in (40) over that in (52)?

The choice is based on the way varactors are characterized experimentally. Capacitance is typically measured by applying a small signal voltage around a constant bias voltage. By varying the bias current, a curve of capacitance versus bias voltage can be obtained. This conforms to the definition of capacitance given in (40). Therefore the only way to map the vendor-provided measurements to a mathematical model is that used in Section 4.

References

[1] A. Hajimiri and T. H. Lee, "A general theory for phase noise in electrical oscillators," *IEEE Journal of Solid-State Circuits*, vol. 33, no. 2, pp. 179-194, Feb. 1998.

[2] J. J. Rael and A. A. Abidi, "Physical processes of phase noise in differential LC oscillators," in *Proc. of the IEEE Custom Integrated Circuits Conference,* pp. 569-572, Orlando, 2000.

[3] E. Hegazi, H. Sjoland and A. A. Abidi, "A filtering technique to lower LC oscillator phase noise," *IEEE Journal of solid-state circuits,* Vol. 36 no. 12, pp. 1921-1930 Dec. 2001.

[4] K. Hoshino, E. Hegazi, J. Rael and A. A. Abidi, "A 1.5 V, 1.7 mA 700 MHz CMOS LC oscillator with no upconverted flicker noise," *Proc. of the European Solid State Circuits Conference*, Villach, 2001.

[5] D. Ham and A. Hajimiri, "Concepts and methods in optimization of integrated LC VCOs," *IEEE Journal of Solid-State Circuits*, Vol. 36 no. 6, pp. 896-909, June 2001.

[6] C. Samori, A. L. Lacaita, F. Villa and F. Zappa, "Spectrum folding and phase noise in LC tuned oscillators," *IEEE Trans on Circuits and Systems II: Analog and Digital Signal Processing*, vol. 45, no. 7, pp. 781-90, 1998.

[7] K. A. Kouznetsov and R. G. Meyer, "Phase noise in LC oscillators," *IEEE Journal of Solid-State Circuits,* Vol. 35 no. 8, pp 1244-1248, Aug. 2000.

[8] C. Samori, A. L. Lacaita, A. Zanchi, S. Levantino and F. Torrisi, "Impact of indirect stability on phase noise performance of fully integrated LC tuned VCOs," in *Proc. Eur. Solid-State Circuits Conf., Germany,* 1999, pp. 202 205.

[9] P. Andreani and S. Mattisson, "On the use of MOS varactors in RF VCOs," *IEEE Journal of Solid-State Circuits,* Vol. 35 no. 6, pp. 905-910, June 2000.

[10] J. Groszkowski, *Frequency of Self-Oscillations*. Oxford: Pergamon Press, 1964.

[11] A. A. Abidi, "How phase noise appears in oscillators," in *Analog Circuit Design: RF A/D Converters, Sensor and Actuator Interfaces, Low-Noise Oscillators, PLLs, and Synthesizers*, R. J. van de Plassche, J. H. Huijsing, and W. Sansen, Eds. Boston, MA: Kluwer, 1997, pp. 428-448.

[12] Yannis Tsividis, *Operation and Model of the MOS Transistor*, second edition, McGraw-Hill, 1998.

[13] A. Kral, F. Behbahani and A. A. Abidi, "RF CMOS oscillators with switched tuning," *Proc. IEEE Custom Integrated Circuits Conf.*, Santa Clara, pp. 555-558, 1998.

[14] Z. Tang, *Private Communication.*

Index

A
absolute jitter 14
amplitude modulation 17
AM-to-FM noise conversion 186

B
Barkhausen criterion 5

C
Colpitts oscillator 67, 89
  current source noise 78
  differential 91
  noise sources 72
  resistor noise 74
  transistor noise 78
complementary differential oscillator 94
current source noise in Colpitts oscillator 78
current-biased differential oscillator 85, 96
cycle-to-cycle jitter 16

D
Demir's model 28
design example
  oscillator for GSM receiver 99, 129
device stress, effect on amplitude 122
differential oscillator 35, 94
  current biased 85, 96
  differential pair noise 50
  resonator noise 44
  tail current noise 57
  voltage biased 88
diffusivity 15

E
example
  oscillator for GSM receiver 99, 129

F
figure of merit 83, 132
FLEX oscillator 168
flicker noise 96, 110, 166
  jitter 15
  Leeson's model 8
  nulling 168, 172
  top-biased oscillator 113
FM noise
  AM-to-FM conversion 186
  current source 155
  due to switch 151
  Groszkowski 142

G
Groszkowski 142
GSM phase noise requirements 99

H
Hajimiri's model 25
harmonic balance 84

I
impulse sensitivity function (ISF) 25

J
jitter 12
  flicker noise 15

K
Kartner 29

L
$\mathcal{L}$ (noise-to-carrier ratio) 8, 19
Leeson's formula ix, 8, 38, 62, 97
linear oscillator 3
linewidth 15, 21
Lorentzian spectrum 23

M
modulation 17

N

noise factor
  Colpitts oscillator 78
  differential oscillator 104
  FET (γ) 61
  Leeson 38, 62, 97
noise filtering 112
  earlier work 125
  wideband 172
noise sensitivity curve 184

O

oscillator
  Colpitts 67, 89
  complementary differential 94
  current-biased differential 85, 96
  design example
    GSM receiver 99, 129
  differential 35
  figure of merit 83, 132
  ideal model 1
  linear 3
  noise filtering 112
  van der Pol 142
  voltage-biased differential 88

P

period jitter 16
phase jitter 15
phase modulation 17
phase noise 19, 104
  Colpitts 72
  flicker noise 139, 166
  FM due to current source 155
  FM due to Groszkowski 142
  FM due to switch 151
  thermal 39
power supply rejection 119

R

resistor noise 6
  in Colpitts oscillator 74
resonator noise in differential oscillator 44

S

single sideband 19
stress, effect on amplitude 122
supply rejection 119
supply sensitivity 187

T

tail current noise in differential oscillator 57
transconductor noise 6
transistor noise in Colpitts oscillator 78
tuning range 123
tuning sensitivity 187

V

van der Pol oscillator 142
varactor 175
  noise sensitivity 181
voltage-biased differential oscillator 88